A Concise Introduction to Linear Algebra

A Concise Introduction to Linear Algebra

Editor

Linsen Chou

scitus
academics

A Concise Introduction to Linear Algebra

Edited by **Linsen Chou**

ISBN: 978-1-68117-179-1

Library of Congress Control Number: 2015949093

SCITUS Academics LLC,
www.scitusacademics.com
Box No. 4766, 616 Corporate Way,
Suite 2, Valley Cottage,
NY 10989

Printed in United States of America on Acid Free Paper ∞

Preface

Linear algebra is the branch of mathematics regarding vector spaces andlinear mappings between such spaces. It includes the study of lines, planes, and subspaces, but is also concerned with properties common to all vector spaces. Linear algebra is the study of linear sets of equations and their transformation properties. Linear algebra allows the analysis of rotations in space, least squares fitting, solution of coupled differential equations, determination of a circle passing through three given points, as well as many other problems in mathematics, physics, and engineering. The matrix and determinant are extremely useful tools of linear algebra. Techniques from linear algebra are also used in analytic geometry, engineering, physics, natural sciences, computer science,computer animation, and the social sciences. Because linear algebra is such a well-developed theory, nonlinear mathematical models are sometimes approximated by linear models. Linear algebra is central to both pure and applied mathematics. Functional analysis studies the infinite-dimensional version of the theory of vector spaces. Combined with calculus, linear algebra facilitates the solution of linear systems of differential equations

This book, A Concise Introduction to Linear Algebra, covers matrix theory and linear algebra, emphasizing topics useful in other disciplines such as physics, economics and social sciences, natural sciences, and engineering.

Table of Contents

A New Iterative Solution Method for Solving Multiple Linear Systems 1

On the FOM Algorithm for the Resolution of the Linear Systems Ax = b 19

Rhotrix Linear Transformation 47

On Solving Centrosymmetric Linear Systems 61

Numerical Quadratures Using the Interpolation Method of Hurwitz-Radon Matrices 91

A Modified Precondition in the Gauss-Seidel Method 109

Solving the Class Equation xd = b in an Alternating Group for Each $\beta \in C^a \cap H_n^c$ and n>1 131

Jordan Semi-Triple Multiplicative Maps on the Symmetric Matrices 153

Generalized Form of Hermite Matrix Polynomials via the Hypergeometric Matrix Function 173

Latent Structure Linear Regression 189

Index 221

A New Iterative Solution Method for Solving Multiple Linear Systems

Saeed Karimi
Department of Mathematics,
Persian Gulf University, Bushehr and Iran

1

Abstract

In this paper, a new iterative solution method is proposed for solving multiple linear systems $A^{(i)}x^{(i)} = b^{(i)}$, for $1 \leq i \leq s$, where the coefficient matrices $A^{(i)}$ and the right-hand sides $b^{(i)}$ are arbitrary in general. The proposed method is based on the global least squares (GL-LSQR) method. A linear operator $\mathcal{L}: \mathbb{R}^{n \times s} \rightarrow \mathbb{R}^{n \times s}$ is defined to connect all the linear systems together. To approximate all numerical solutions of the multiple linear systems simultaneously, the GL-LSQR method is applied for the operator and the approximate solutions are obtained recursively. The presented method is compared with the well-known LSQR method. Finally, numerical experiments on test matrices are presented to show the efficiency of the new method.

Introduction

We want to solve, using global least squares (GL-LSQR) method, the following linear systems:

$$A^{(i)}x^{(i)} = b^{(i)}, \quad 1 \leq i \leq s \tag{1}$$

Where $A^{(i)}$ are arbitrary matrices of order n, and in general $A^{(i)} \neq A^{(j)}$ and $b^{(i)} \neq b^{(j)}$ for $i \neq j$ In spite of that, in many practical application the coefficient matrices and the right-hand sides are not arbitrary, and often there is information that can be shared among the coefficient matrices and right-hand sides. Multiple linear systems (1) arise in many problems in scientific computing and engineering application, including recursive least squares computations [1, 2], wave scattering problems [3, 4], numerical methods for integral equations [4, 5], and image restorations [6].

Many authors, see [7, 8], have researched to approximate the solutions of the multiple linear systems (1) with the same coefficient matrix but different right-hand sides, i.e

$$Ax^{(i)} = b^{(i)}, \qquad 1 \leq i \leq s. \tag{2}$$

Recently, S. Karimi and F. Toutounian proposed an iterative method for solving the linear systems (2) with the some advantages over the existing methods, see [9–11], for more details. In [12], Tony F. chan and Michael K. Ng presented the Galerkin projection method for solving linear systems (1). They focused on the seed projection method which generates a Krylov subspace from a set of direction vectors obtained by solving one of the systems, called the seed system, by the CG method and then projects the residuals of other systems onto the generated Krylov subspace to get the approximate solutions. Note that the coefficient matrices in this method are real symmetric positive definite.

In this paper, we propose a new method to solve the linear systems (1) simultaneously, where the coefficient matrices and right-hand sides are arbitrary. We define a linear operator $\mathcal{L}: \mathbb{R}^{n \times s} \rightarrow \mathbb{R}^{n \times s}$ to connect all the linear systems (1) together. Then we apply the GL-LSQR method [13] for the linear operator $\mathcal{L}$ and obtain recursively the approximate solutions simultaneously. In the new method, the linear operator $\mathcal{L}$ will be reduced to a lower global bidiagonal, namely $\mathcal{L}$-Bidiag, matrix form. We obtain a recurrence formula for generating the sequence of approximate solutions. Our new method has certain advantages over the existent methods. In the new method, the coefficient matrices $A^{(i)}$ and right-hand sides $b^{(i)}$ are arbitrary. Also we do not need to store the basis vectors, we do not need to predetermine a subspace dimension and the approximate solutions and residuals are cheaply computed at every stage of the algorithm because they are updated with short-term recurrence.

The remainder of the paper is organized as follows. Section 2 is devoted to a short review of the global least squares (GL-LSQR) method. In section 3, we present a new method, namely $\mathcal{L}$-GL-LSQR, to solve the multiple linear systems (1). We also show how to reduce the low-rank approximate solutions to linear operator $\mathcal{L}$. The section 4 is devoted to some numerical experiments and comparing the new method with the well-known LSQR method, when it is applied to s linear systems independently. Finally, we make some concluding remarks in section 5.

We use the following notations. For X and Y two matrices in $\mathbb{R}^{n\times s}$, we consider the following inner product $\langle X, Y \rangle_F = tr(X^T Y)$, where tr(.) denotes the trace of a matrix. The associated norm is the Frobenious norm denoted by $\|.\|_F$. The notation $X \perp_F Y$ means that $\langle X, Y \rangle_F = 0$ and X (:, i) means that the ith column of X. Finally, we use the notation * for the following product:

$$\mathcal{V} * y = \sum_{j=1}^{m} y_j V_j, \tag{3}$$

Where $\mathcal{V} = [V_1, V_2, \cdots V_m], V_j \in \mathbb{R}^{n\times s}$ for $1 \le j \le m$, and $y \in \mathbb{R}^m$.
By the same way, we define

$$\mathcal{V} * T = [\mathcal{V} * T(:,1), \mathcal{V} * T(:,2), \cdots, \mathcal{V} * T(:,m)], \tag{4}$$

Where T is the m×m matrix. It is easy to show that the following relations are satisfied:

$$\mathcal{V} * (y + z) = \mathcal{V} * y + \mathcal{V} * z, \quad (\mathcal{V} * T) * y = \mathcal{V} * (Ty), \tag{5}$$

Where y and z are two vectors of $\mathbb{R}^m$.

THE GLOBAL LEAST SQUARES (GL-LSQR) METHOD

In this section, we recall some fundamental properties of GL-LSQR algorithm [13], which is an iterative method for solving the multiple linear systems (2). When all the $b^{(i)}$'s are available simultaneously, the multiple linear systems (2) can be written as

$$AX = B, \tag{6}$$

Where A is an n×n nonsingular arbitrary matrix, B and X are an n×s rectangular matrices whose columns are $b^{(1)}, b^{(2)}, \cdots, b^{(s)}$ and $x^{(1)}, x^{(2)}, \cdots, x^{(s)}$, respectively.

For solving the matrix Equation (6), the GL-LSQR method uses a procedure, namely Global-Bidiag procedure, to reduce A to the global lower bidiagonal form. The Global-Bidiag procedure can be described as follows.

Global-Bidiag (starting matrix B; reduction to global lower bidiagonal form):

$$\beta_1 U_1 = B, \alpha_1 V_1 = A^T U_1$$

$$\left.\begin{array}{l} \beta_{i+1} U_{i+1} = A V_i - \alpha_i U_i \\ \alpha_{i+1} V_{i+1} = A^T U_{i+1} - \beta_{i+1} V_i \end{array}\right\}, \quad i = 1, 2, \cdots, \tag{7}$$

Where $U_i, V_i \in \mathbb{R}^{n \times s}$. The scalars $\alpha_i \geq 0$ and $\beta_i \geq 0$ are chosen so that $\|U_i\|_F = \|V_i\|_F = 1$.

With the definitions the recurrence relations (7) may be rewritten as:

$$\mathcal{U}_k \equiv [U_1, U_2, \cdots, U_k],$$

$$\mathcal{V}_k \equiv [V_1, V_2, \cdots, V_k],$$

$$T_k \equiv \begin{bmatrix} \alpha_1 & & & \\ \beta_2 & \alpha_2 & & \\ & \ddots & \ddots & \\ & & \beta_k & \alpha_k \\ & & & \beta_{k+1} \end{bmatrix},$$

$$\mathcal{U}_{k+1} * (\beta_1 e_1) = B, \tag{8}$$

$$A\mathcal{V}_k = \mathcal{U}_{k+1} * T_k, \tag{9}$$

$$A^T \mathcal{U}_{k+1} = \mathcal{V}_k * T_k^T + \alpha_{k+1} V_{k+1} * e_{k+1}^T. \tag{10}$$

For the Global-Bidiag procedure, we have the following propositions.

Proposition 1 [13]: Suppose that k step of the GlobalBidiag procedure have been taken, then the n×s block vectors $V_1, V_2, \ldots, V_k$ and $U_1, U_2, \ldots, U_k, U_{k+1}$ are F orthonormal basis of the Krylov subspaces $\mathcal{K}_k(A^T A, V_1)$ and $\mathcal{K}_{k+1}(AA^T, U_1)$, respectively.

Proposition 2 [13]: The Global-Bidiag procedure will be stopped at step m if and only if, m=min {μ, λ} where μ and λ are the grades of V_1 and U_1 with respect to $A^T A$ and AA^T, respectively.

Proposition 3 [13]: Let u_k be the matrix defined by $u_k \equiv [U_1, U_2, \ldots, U_k]$ where the n×s matrices U_i, i=1, ,k, are generated by the Global-Bidiag procedure. Then

$$\|u_k * \eta\|_F = \|\eta\|_2 .$$

By using the Global-Bidiag procedure, the GL-LSQR algorithm constructs an approximate solution of the form

$X_k = V_k * y_k$, where $y_k = \left[y_k^{(1)}, \cdots, y_k^{(k)}\right]^T \in \mathbb{R}^k$, which solves the least-squares problem,

$$\min_X \|B - AX\|_F .$$

The main steps of the GL-LSQR algorithm can be summarized as follows.

Algorithm 1: GI-LSQR algorithm

1. Set $X_0 = 0$
2. $\beta_1 = \|B\|_F$, $U_1 = B/\beta_1$, $\alpha_1 = \|A^T U_1\|_F$, $V_1 = A^T U_1/\alpha_1$.
3. Set $W_1 = V_1$, $\overline{\phi}_1 = \beta_1$, $\overline{\rho}_1 = \alpha_1$
4. For i=1,2 until convergence, Do:
5. $\overline{W}_i = AV_i - \alpha_i U_i$
6. $\beta_{i+1} = \|\overline{W}_i\|_F$
7. $U_{i+1} = \overline{W}_i/\beta_{i+1}$
8. $\overline{S}_i = A^T U_{i+1} - \beta_{i+1} V_i$
9. $\alpha_{i+1} = \|\overline{S}_i\|_F$
10. $V_{i+1} = \overline{S}_i/\alpha_{i+1}$
11. $\rho_i = \left(\overline{\rho}_i^2 + \beta_{i+1}^2\right)^{1/2}$
12. $c_i = \overline{\rho}_i/\rho$
13. $s_i = \beta_{i+1}/\rho_i$
14. $\theta_{i+1} = s_i \alpha_{i+1}$
15. $\overline{\rho}_{i+1} = c_i \alpha_{i+1}$
16. $\phi_i = a_i \overline{\phi}_i$
17. $\phi_{i+1} = c_i \overline{\phi}_i$
18. $\phi_i = c_i \overline{\phi}_i$
19. $\phi_{i+1} = -s_i \phi_i$

20. $X_i = X_{i-1} + (\phi_i / \rho_i) W_i$

21. $W_{i+1} = V_{i+1} - (\theta_{i+1} / \rho_i) W_i$

22.If $|\bar{\phi}_{i+1}|$ is small enough then stop 23) End Do.

More details about the GL-LSQR algorithm can be found [13].

THE GL-LSQR-LIKE OPERATOR METHOD

In this section, we propose a new method for solving the linear systems (1). For this mean, we define the following linear operator

$$\mathcal{L} : \mathbb{R}^{n \times s} \to \mathbb{R}^{n \times s} \tag{11}$$

$$\mathcal{L}(X) = \left[A^{(1)} X(:,1), \cdots, A^{(s)} X(:,s)\right], \tag{12}$$

Where $A^{(j)}$, J=1, , s are the coefficient matrices of the multiple linear systems (1). Therefore, the linear systems (1) is written as:

$$\mathcal{L}(X) = B, \tag{13}$$

Where B is an n×s rectangular matrix whose columns are $b^{(1)}, b^{(2)}, \cdots, b^{(s)}$ the right hand sides of the linear systems (1).

Definition 1: Let $\mathcal{L}$ be linear operator (11). Then

$$\mathcal{L}^T X = [A^{(1)T} X(:,1), \cdots, A^{(s)T} X(:,S)]$$

Consider the block operator (13), similar to the wellknown block Krylov subspace

$$\mathcal{K}_m(A;R) = \text{span}\{R, AR, A^2R, \cdots, A^{m-1}R\}$$

Where R is the residual of the operator equation (13), we define the following block Krylov-like subspace.

Definition 2: Let $\mathcal{L}$ be linear operator (11). Then

$$\mathcal{K}_m(\mathcal{L};R) = \text{span}\{R, \mathcal{L}(R), \cdots, \mathcal{L}^{m-1}(R)\}$$

Where $\mathcal{L}^i = \overset{i\,\text{time}}{\overline{\mathcal{L} o \mathcal{L} o \ldots o \mathcal{L}}}$ and o is the combination of two operators.

Definition 3: Let $\mathcal{L}$ be linear operator (11) and $\nu_m = [V_1, V_2, \cdots, V_m] \in \mathbb{R}^{n\times s}$. Then

$$\mathcal{L}(\nu_m) = [\mathcal{L}(V_1), \mathcal{L}(V_2), \cdots, \mathcal{L}(V_m)] \in \mathbb{R}^{n\times ms}$$

Where $V_j \in \mathbb{R}^{n\times s}$, $j = 1, 2\cdots, s$.

To approximate the solution of the block operator Equation (13), we present a new algorithm, will be referred to $\mathcal{L}$-GL-LSQR algorithm, which is based on the Global-Bidiag-like procedure, will be referred to $\mathcal{L}$ –Bidiag. The $\mathcal{L}$-Bidig procedure reduces the linear operator $\mathcal{L}$ to the lower bidiagonal matrix form. This procedure can be described as follows.

$\mathcal{L}$-Bidiag (starting matrix B; reduction to lower bidiagonal matrix form):

$$\beta_1 U_1 = B, \alpha_1 V_1 = \mathcal{L}^T(U_1)$$

$$\left.\begin{aligned} \beta_{i+1} U_{i+1} &= \mathcal{L}(V_i) - \alpha_i U_i \\ \alpha_{i+1} V_{i+1} &= \mathcal{L}^T(U_{i+1}) - \beta_{i+1} V_i \end{aligned}\right\}, \quad i = 1, 2, \cdots, \tag{14}$$

Where $U_i, V_i \in \mathbb{R}^{n\times s}$. The scalars $\alpha_1 \geq 0$ and $\beta_1 \geq 0$ are chosen so that $\|U_i\|_F = \|V_i\|_F = 1$.

Similar to Global-Bidiag procedure, we define

$$\mathcal{U}_m \equiv [U_1, U_2, \cdots, U_m],$$
$$\mathcal{V}_m \equiv [V_1, V_2, \cdots, V_m],$$
$$T_m \equiv \begin{bmatrix} \alpha_1 & & & \\ \beta_2 & \alpha_2 & & \\ & \ddots & \ddots & \\ & & \beta_m & \alpha_m \\ & & & \beta_{m+1} \end{bmatrix}.$$

According to notation * and by using the definition 3, the recurrence relations (14) may be rewritten as:

$$\mathcal{U}_{m+1} * (\beta_1 e_1) = B, \tag{15}$$

$$\mathcal{L}(\mathcal{V}_m) = \mathcal{U}_{m+1} * T_m, \tag{16}$$

$$\mathcal{L}^T(\mathcal{U}_{m+1}) = \mathcal{V}_m * T_k^T + \alpha_{m+1} V_{m+1} * e_{m+1}^T. \tag{17}$$

Proposition 4: Suppose that k step of the $\mathcal{L}$-Bidiag procedure have been taken, then the block vectors $V_1, \cdots, V_m$ and $U_1, \cdots U_{m+1}$ are F-orthonormal basis of the Krylov-like subspace $\mathcal{K}_m(\mathcal{L}^T \mathcal{L}; V_1)$ and $\mathcal{K}_{m+1}(\mathcal{L}\mathcal{L}^T; U_1)$, respectively.

The proof of this proposition is similar to that given in [13].

The quantities generated from the linear operator $\mathcal{L}$ and B by the $\mathcal{L}$-Bidiag process will now be used to solve the block least squares problem,

$$\min_X \|B - \mathcal{L}(X)\|_F.$$

Let the quantities

$$X_m = \mathcal{V}_m * y_m, \tag{18}$$

$$R_m = B - \mathcal{L}(X_m), \tag{19}$$

be defined, where $\in \mathbb{R}$. According to linearity of the operator, it is easy to show that $\mathcal{L}(X_m) = \mathcal{L}(\mathcal{V}_m) * y_m$.

Also it readily follows from (15), (16) and properties of product * the equation holds to working accuracy.

$$\begin{aligned} R_m &= B - \mathcal{L}(\mathcal{V}_m) * y_m \\ &= \mathcal{U}_{m+1} * (\beta_1 e_1) - (\mathcal{U}_{m+1} * T_m) * y_m \\ &= \mathcal{U}_{m+1} * (\beta_1 e_1 - T_m y_m), \end{aligned}$$

To minimize the mth residual $\|R_m\|_F$, since $\mathcal{U}_{m+1}$ is F-orthonormal and by using the proposition 3, we choose y_m so that is minimum.

$$\|R_m\|_F = \|\beta_1 e_1 - T_m y_m\|_2, \tag{20}$$

This minimization problem is carried out by applying the QR decomposition [13], where a unitary matrix Q_m is determined so that

$$Q_m \begin{bmatrix} T_m & \beta_1 e_1 \end{bmatrix} = \begin{bmatrix} \mathcal{R}_m & f_m \\ 0 & \bar{\phi}_{m+1} \end{bmatrix}$$

$$= \begin{bmatrix} \rho_1 & \theta_1 & & & & \phi_1 \\ & \rho_2 & \theta_3 & & & \phi_2 \\ & & \ddots & \ddots & & \vdots \\ & & & \rho_{m-1} & \theta_m & \phi_{m-1} \\ & & & & \rho_m & \phi_m \\ & & & & 0 & \phi_{m+1} \end{bmatrix},$$

Where ρ_1 θ_1 and ϕ_1 are scalars. The above QR factorization is determined by constructing the mth plane rotation $Q_{m,m+1}$ to operate on rows m and m+1 of the transformed $[T_m\ \beta_1 e_1]$ to annihilate β_{m+1}. This gives the following simple recurrence relation:

$$\begin{bmatrix} c_m & s_m \\ -s_m & c_m \end{bmatrix}\begin{bmatrix} \bar{\rho}_m & 0 & \bar{\phi}_m \\ \beta_{m+1} & \alpha_{m+1} & 0 \end{bmatrix} = \begin{bmatrix} \rho_m & \theta_{m+1} & \phi_m \\ 0 & \bar{\rho}_{m+1} & \bar{\phi}_{m+1} \end{bmatrix},$$

Where $\bar{\rho}_1 \equiv \alpha_1, \bar{\phi}_1 \equiv \beta_1$ and the scalars c_m and s_m are the nontrivial elements of $Q_{m,m+1}$ the quantity $\bar{\rho}_m$ and $\bar{\phi}_m$ are intermediate scalars that are subsequently replaced by ρ_m and ϕ_m

With setting

$$y_m = \mathcal{R}_m^{-1} * f_m,$$

the approximate solution is given by

$$X_m = \mathcal{V}_m * \left(\mathcal{R}_m^{-1} * y_m\right), \tag{21}$$

$$= \left(\mathcal{V}_m * \mathcal{R}_m^{-1}\right) * f_m. \tag{22}$$

Letting

$$\mathcal{P}_m \equiv \mathcal{V}_m * \mathcal{R}_m^{-1} \equiv \left[P_1 P_2 \cdots P_m\right],$$

then

$$X_m = \mathcal{P}_m * f_m.$$

The n×s matrix P_m the last block column of $\mathcal{P}_m$, can be computed from the previous P_{m-1} and V_m, by the simple update

$$P_m = (V_m - \theta_m P_{m-1})/\rho_m, \tag{23}$$

also note that,

$$f_m = \begin{bmatrix} f_{m-1} \\ \phi_m \end{bmatrix},$$

in which

$$\phi_m = c_m \bar{\phi}_m.$$

Thus, X_m can be updated at each step, via the relation

$$X_m = X_{m-1} + \phi_m P_m.$$

The residual norm $\|R_m\|_F$ is computed directly from the quantity $\bar{\phi}_{m+1}$ as

$$\|R_m\|_F = |\bar{\phi}_{m+1}|.$$

Some of the work in (23) can be eliminated by using matrices $W_m = \rho_m P_m$ in place of P_m. The main steps of the $\mathcal{L}$-GL-LSQR algorithm can be summarized as follows.

Algorithm 2: $\mathcal{L}$-GL-LSQR algorithm

1. Set $X_0 = 0$
2. $\beta_1 = \|B\|_F$, $U_1 = B/\beta_1$, $\alpha_1 = \|\mathcal{L}^T(U_1)\|_F$, $V_1 = \mathcal{L}^T(U_1)/\alpha_1$.
3. Set, $W_1 = V_1$, $\bar{\phi}_1 = \beta_1$, $\bar{\rho}_1 = \alpha_1$,

4. For $i = 2, 2, \cdots$ until convergence, Do:
5. $\hat{W}_i = \mathcal{L}(V_i) - \alpha_i U_i$
6. $\beta_{i+1} = \left\| \hat{W}_i \right\|_F$
7. $U_{i+1} = \hat{W}_i / \beta_{i+1}$
8. $\hat{S}_i = \mathcal{L}^T(U_{i+1}) - \beta_{i+1} V_i$
9. $\alpha_{i+1} = \left\| \hat{S}_i \right\|_F$
10. $V_{i+1} = \hat{S}_i / \alpha_{i+1}$
11. $\rho_i = \left(\overline{\rho}_i^2 + \beta_{i+1}^2 \right)^{1/2}$
12. $c_i = \overline{\rho}_i / \rho$
13. $s_i = \beta_{i+1} / \rho_i$
14. $\theta_{i+1} = s_i \alpha_{i+1}$
15. $\overline{\rho}_{i+1} = c_i \alpha_{i+1}$
16. $\phi_i = a_i \overline{\phi}_i$
17. $\overline{\phi}_{i+1} = c_i \overline{\phi}_i$
18. $\phi_i = c_i \overline{\phi}_i$
19. $\overline{\phi}_{i+1} = -s_i \phi_i$
20. $X_i = X_{i-1} + (\phi_i / \rho_i) W_i$
21. $W_{i+1} = V_{i+1} - (\theta_{i+1} / \rho_i) W_i$
22. If $|\overline{\phi}_{i+1}|$ is small enough then stop 23) EndDo.

As we observe, the $\mathcal{L}$-GL-LSQR algorithm has certain advantages, we obtain simultaneously the approximate solution of the multiple linear systems (1). Also the residual norm is cheaply computed at every stage of the algorithm.

As a application of the new method applying it to solve the Sylvester equation in a special case. Consider the Sylvester equation

$$AX - XB = C,$$

Where $A \in \mathbb{R}^{n\times n}$, $C \in \mathbb{R}^{n\times s}$ and $B \in \mathbb{R}^{s\times s}$ are known and $X \in \mathbb{R}^{n\times s}$ is unknown. By using the Schur decomposition for the symmetric matrix B, the above Sylvester equation is returned to the s linear systems as follows. There is a unitary matrix Q such that $B = Q\Lambda Q^T$, where Λ; is a diagonal matrix which diagonal elements are the eigenvalues of B. So we have

$$AXQ - XE\Lambda = CQ = \hat{C},$$

by taking $\hat{X} = XQ$ the Sylvester equation is converted to the following s linear systems

$$\hat{A}^{(j)}\hat{x}^{(j)} = \hat{c}^{(j)},\ j = i,\cdots,s,$$

Where $\hat{A}^{(j)} = A - \lambda_j I$, $\hat{x}^{(j)}$ and $\hat{c}^{(j)}$ are the j-th colmun of $\hat{X}$ and $\hat{C}$, respectively.

NUMERICAL EXPERIMENTS

In this section, all the numerical experiments were computed in double precision with some MATLAB codes. For all the examples the initial guess X_0 was taken to be zero matrix. We consider two sets of numerical experiments. The first set of numerical experiments contains matrices of the form

$$A^{(i)} = A - \lambda_i I,\ i = 1,2,\cdots,s,$$

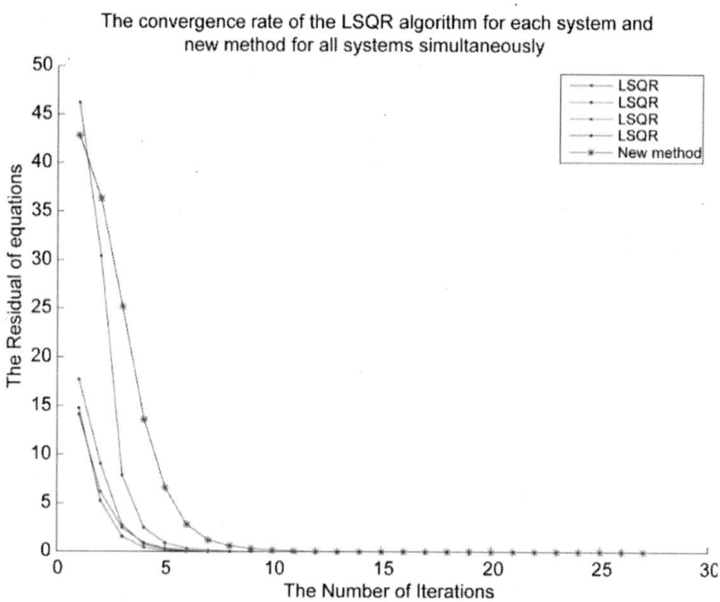

Figure 1: Convergence history of the LSQR algorithm and the new algorithm for the first set matrices with s = 4 and n = 3000.

obtained from the Sylvester equation, explained in previous section, where $A = \text{tradiag}\left(-1+\frac{10}{n+1}, 4, -1-\frac{10}{n+1}\right)$ and λ_i is the eigenvalue of the matrix $B = \text{tridiag}\left(-1+\frac{1}{s+1}, 2, -1+\frac{1}{s+1}\right)$. The right-hand side matrix C is taken rand(n,s), where the function rand creates an n×s random matrix. The matrices of the second set of experiments arise from the three-point centered discretization of the operator $-\frac{d}{dx}\left(a(x)\frac{du}{dx}\right)$ in [0, 1] where the function a(x) is given by a(x) = c + dx, where c and d are two parameters. The discretization is performed using a grid size of $h=\frac{1}{65}$, yielding matrices of size 64 with the values of $c_k = 0.1551\times(0.9524)^k$ and $d_k = 7.7566\times(0.9524)^k, k = 1,2,\cdots,s$. The right hand sides of these systems are generated randomly with their 2-norms being 1. All the tests were stopped as soon as, $|\bar{\phi}| \le 10^{-7}$.

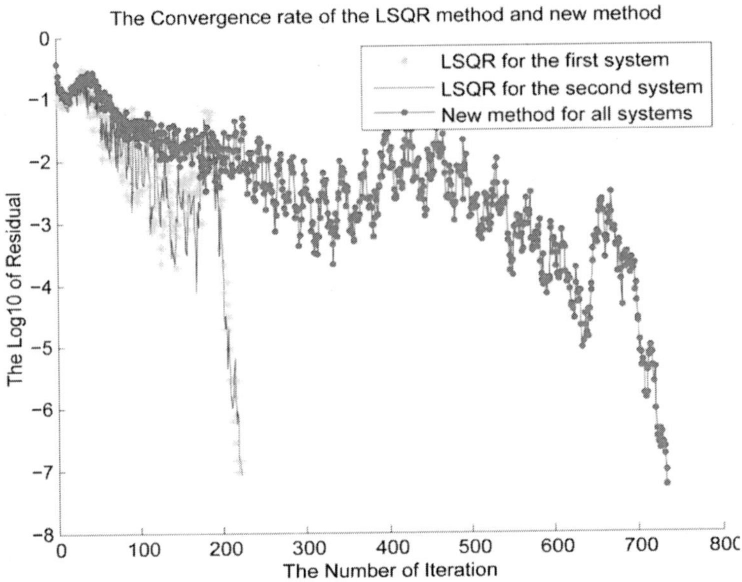

Figure 2: Convergence history of the LSQR algorithm and the new algorithm for the second set matrices with s = 2.

We display the convergence history in Figure 1 and Figure 2 for the systems corresponding to the matrices of the first set of matrices and second set of matrices, respectively. Figure 1 shows that the $\mathcal{L}$-GL-LSQR algorithm converges, however, Figure 2 shows slowly convergence which can be remedied by using a reliable preconditioner. But we did not deal with the preconditioner techniques in this paper.

CONCLUSIONS

We proposed a new method for solving multiple linear systems $A^{(i)}x^{(i)} = b^{(i)}$, for $1 \le i \le s$, where the coefficient matrices $A^{(i)}$ and the right-hand sides $b^{(i)}$ are arbitrary in general. This method has certain advantages which is applied to the arbitrary coefficient matrices $A^{(i)}$ and right-hand sides $b^{(i)}$. Also it is not needed to store the basis vectors, it is not needed to predetermine a subspace dimension and the approximate solutions and residuals are cheaply computed at every stage of the algorithm simultaneously because they are updated with short-term recurrence. Applying a reliable preconditioner for the linear systems of Equation (1) may increas the convergence rate, which has not been dicussing in this paper.

ACKNOWLEDGEMENTS

The author would like to thank the anonymous referees for their helpful suggestions, which would greatly inprove the article.

This work is financially supported by the research committee of Persian Gulf University.

REFERENCES

1. C. C. Paige and M. A. Saunders, "LSQR: An Algorithm for Sparse Linear Equations and Sparse Least Squares," ACM Transactions on Mathematical, Vol. 8, No. 1, 1982, pp. 43-71. doi:10.1145/355984.355989
2. R. Plemmons, "FFT-Based RLS in Signal Processing," Proceeding of the IEEE International Conference on Acoustics, Speech and Signal Processing, Minneapolis, 27-30 April 1993, pp. 571-574.
3. W. E. Boyse and A. A. Seidl, "A Block QMR Method for Computing Multiple Simultaneous Solutions to Compex Symmetric Systems," SIAM Journal on Scientific Computing, Vol. 17, No. 1, 1996, pp. 263-274.
4. R. Kress, "Linear Integral Equations," Springer-Verlag, New York, 1989. doi:10.1007/978-3-642-97146-4
5. D. O'Leary, "The Block Conjugate Gradient Algorithm and Related Methods," Linear Algebra and Its Applications, Vol. 29, 1980, pp. 293-322. doi:10.1016/0024-3795(80)90247-5
6. A. Jain, "Fundamentals of Digital Image Processing," Prentice-Hall, Englewwood Cliffs, 1989.
7. S. Karimi and F. Toutounian, "The block Least Squares Method for Solving Nonsymmetric Linear Systems with Multiple Right-Hand Sides," Applied Mathematics and Computation, Vol. 177, No. 2, 2006, pp. 852-862. doi:10.1016/j.amc.2005.11.038
8. G. Golub and C. Loan, "Matrix Computations," 2nd Edition, Johns Hopkins Press, Bultimore, 1989.
9. A. El Guennouni, K. Jbilou and H. Sadok, "The Block Lanczos Method for Linear Systems with Multiple RightHand Sides," Applied Numerical Mathematics, Vol. 51, No. 2-3, 2004, pp. 243-256.

10. A. El Guennouni, K. Jbilou and H. Sadok, "A Block BiCGSTAB Algorithm for Mutiple Linear Systems," Electronic Transactions on Numerical Analysis, Vol. 16, 2003, pp. 129-142.
11. S. Karimi and F. Toutounian, "On the Convergence of the BL-LSQR Algorithm for Solving Matrix Equations," International Journal of Computer Mathematics, Vol. 88, No. 1, 2011, pp. 171-182. doi:10.1080/00207160903365883
12. T. F. Chan and M. K. Ng, "Galerkin Projection Methods for Solving Multiple Linear Systems," SIAM Journal on Scientific Computing, Vol. 21, No. 3, 2012, pp. 836-850.doi:10.1137/S1064827598310227
13. F. Toutounian and S. Karimi, "Global Least Squares (GLLSQR) Method for Solving General Linear Systems with Several Right-Hand Sides," Applied Mathematics and Computation, Vol. 178, No. 2, 2006, pp. 452-460. doi:10.1016/j.amc.2005.11.065.

CITATION

1. S. Karimi, "A New Iterative Solution Method for Solving Multiple Linear Systems," Advances in Linear Algebra & Matrix Theory, Vol. 2 No. 3, 2012, pp. 25-30. doi: 10.4236/alamt.2012.23004.

On the FOM Algorithm for the Resolution of the Linear Systems Ax = b

Mongi Benhamadou
Department of Mathematics, Faculty of Sciences of Sfax, Sfax, Tunisia

2

ABSTRACT

In this paper, we propose another version of the full orthogonalization method (FOM) for the resolution of linear system $Ax = b$, based on an extended definition of Sturm sequence in the calculation of the determinant of an upper hessenberg matrix in $o(n^2)$. We will also give a new version of Givens method based on using a tensor product and matrix addition. This version can be used in parallel calculation.

INTRODUCTION

The resolution of linear systems $Ax = b$ is in the heart of numerous scientific applications: discretization of partiel differentiel equations, image processing, the linearization of nonlinear problems in a sequence of linear problems [1] [2] [3] etc. There are many kinds of methods that resolve linear systems. Some are direct, others are iterative. Generally, direct methods [4] –[6], such as Gauss, Cholesky, QR, are efficient and proved solvers of small size systems. But for problems of big size, these methods require quite prohibitive memory space and therefore become numerically expensive. More and more, they are replaced by iterative methods. Most iterative methods [2] [6] [7] treat linear systems through vector-matrix products with an appropriate data structure permitting the exploitation of sparsity of the matrix A by storing only nonzero elements which are indi–spensable for the calculation of these products, which reduces both the memory size and the processing time. We distinguish two kinds of iterative methods for the resolution of linear systems:

- The asymptotic methods [6]–[8] are older and simpler to implement. Among the most known ones, we mention Jacobi's, Gauss-Seidel's and the relaxation methods. Generally, these methods remain less reliable than the direct methods, and more or less efficient with some specific problems. Currently, these methods are no longer used as linear solvers but remain interesting as preconditioners for other iterative methods. The increase of the size of the systems to be solved leads to the projection methods.
- ***Krylov Methods***: They are being used for some time and proved to be very efficient [2]. Their analysis is more complicated than that of the asymptotic methods. They are based on a technique of projection on a Krylov

subspace, having the form:; K_m (A, r_0) = span (r_0, Ar_0, , $A^{m-1}r_0$); dim (K_m (A, r_0)) = m, spanned by the successive powers of A applied to the risudual vector $r_0 = b - Ax_0$, where x_0 is a vector of $\mathbb{R}^n$, these Krylov's subspaces constitute an increasing sequence of subspaces, one being included in the other and permitting the construction of a sequence of vectors converging towards the solution. The convergence of these methods is theorically assured for m=n maximum. The Krylov methods resolve linear systems having nonsymmetric matrices. The most used Krylov methods based on Arnoldi algorithm are: the full orthogonalization method (FOM) and its varieties: the method of minimal residue (or shortly GMRES), Lanczos method, and that of conjugate gradient for symmetric and symmetric positive definite matrices. These projection methods are more generalisable, more powerful and currently the most used.

In the first section, we will review the orthogonal projection methods on Krylov's subspaces, focusing on the FOM method for the resolution of linear systems.

In the second section, we will extend the definition of Sturm sequence, which will serve for the calculation of the determinant of upper hessenberg matrices.

In the third section, we will give a new variety of Givens method [4] for the resolution of linear systems having an upper hessenberg matrix.

Finally, in the fourth section, we report the results of some numerical tests.

In what follows, we consider a real regular square matrix A, a vector $b \in \mathbb{R}^n$ and the correspondent linear system:

$$Ax = b \tag{1}$$

If x_0 is a starting approximation of solution (1), then the residual vector associated to x_0 is $r_0 = b - Ax_0$. We call Krylov space of order m, noted k_m, the vectorial space spanned by r_0 and its (m–1) iterated by A:

$$K_m\left(A, r_0\right) = vect\left(r_0, Ar_0, \cdots, A^{m-1}r_0\right). \tag{2}$$

The Projection Method

In this part, we will review generallities over iterative methods of projection for the resolution of linear systems (1), by using projections in particular subspaces, namely Krylov's spaces [1] [2].

Most actual iterative techniques used in solving large linear systems use in one way or another a projection procedure. The main idea of projection techniques consist in extracting an approximate solution to system (1) of a subspace of $\mathbb{R}^n$. It leads to a small linear system. This is a basic projection step. If we take K_m (A, r_0) of m dimension (the subspace searched for), then, generally, m constraints must be imposed in order to be able to extract such an approximation. One usual way of describing these constraints is to impose to the residual vectors r = b – Ax to be orthogonal to the linearly independent m vectors, this will define another subspace noted L_m with dimension m, which we call "constraints space". We are then seeking an approximate solution x_m to problem (1) by imposing that $x_m \in K_m$ and that the new residual vector r_m=b-Ax_m must be orthogonal to L_m. If, in addition, we want to use the starting approximation x_0, then the approximate solution x_m must be searched for in the affine space x_0+K_m, then the solution of (1) will be characterized as follows:

$$\begin{cases} x_0 : \text{data} \\ \text{find } x_m \in x_0 + K_m \text{ such that :} \\ r_m \perp L_m. \end{cases} \tag{3}$$

As such, the definition of the projection method is based on two conditions:

- The first fixes the "place" where we should look for the solution at the kth iteration: usually, this place is an affine subvariety of $\mathbb{R}^n$. According to Krylov's methods, this affine subvariety is x_0+K_k (A, r_0), which gives us the following "space condition":

$$x_k \in x_0 + K_k\left(A, r_0\right) \tag{4}$$

- The second condition must fix the x_k that is appropriate to that space (4). We hope that x_k minimizes the vector error $e_k = x - x_k$, or the residual vector r_k = b-Ax_k or one or those two vectors be orthogonal to a space L_k of dimension k. This condition is called "Petrov-Galerkin condition", and is expressed as follows:

$$r_k \perp K_k\left(A, r_0\right). \tag{5}$$

As for the projection method, we will study in the next part, Arnoldi's method for the resolution of linear systems. Let's start now by recalling Arnoldi's algorithm.

Arnoldi's Algorithm

Arnoldi's Algorithm is an iterative process consisting in calculating, simultaneously, an orthonormal basis V_m of K_m, and an upper hessenberg matrix H_m. The matrix H_m is a representation of A with respect to this basis.

In practice, constructing Krylov's spaces leads to determining basis. The natural basis

$$\left\{r_0, Ar_0, \cdots, A^{k-1}r_0\right\}$$

can never be used, due to its numerical degeneration, and the Krylov's matrices become increasingly ill-conditioned [2]. To avoid this numerical degeneration of the natural basis of Krylov's space, the solution consists in putting in place an orthonormalization process. Arnoldi's basis is then constructed by applying the modified Gram-Schmidt orthonormalization method to vectors obtained by successive products matrix vector for euclidien inner product. The Arnoldi's algorithm takes the following form:

$$
(A_{Am})\begin{cases}
\text{data}: n,\ A \in \mathbb{R}^{n\times n},\ m \le n,\ V_m \in \mathbb{R}^{n\times m} \\
x_0 \in^n,\ r_0 = b - Ax_0,\ h_{jp1j} = 1.0,\ j = 1 \\
\bullet\ \text{compute}: v_1 = \dfrac{r_0}{\|r_0\|_2},\ \varepsilon = 10^{-9},\ V_1 = \{v_1\}. \\
\quad \text{while}\left(h_{jp1j} \ge \varepsilon\right) \\
\quad\quad \bullet\ \text{compute}: w = Av_j \\
\quad\quad \text{for } i = 1 \text{ to } j \\
\quad\quad\quad \bullet\ \text{compute}: h_{ij} = \langle w, v_i \rangle \\
\quad\quad\quad \bullet\ \text{compute}: w = w - h_{ij}v_i \\
\quad\quad \text{end } i \\
\quad\quad \bullet\ \text{compute}: h_{j+1,j} = \|w\|_2 \\
\quad\quad \text{if } h_{j+1,j} = 0,\ \text{goto}: \text{"end while"} \\
\quad\quad \bullet\ \text{compute}: v_{j+1} = \dfrac{w}{h_{j+1,j}},\ V_{j+1} := \{v_1, \cdots, v_{j+1}\}. \\
\quad\quad j = j + 1 \\
\quad \text{end while} \\
//\ \text{output}: m = j,\ \text{matrices}:\ V_m, H_m.
\end{cases}
$$

If we assign h_{ij} as coefficient of orthogonalization of A_{vj} with respect to v_i, i =1, 2, j+1 and $h_{j+1,j}$ as norm of vector w obtained by orthogonalization of vector Av_j with respect to vectors v_i, then, the formula of Arnoldi's basis construction can be written:

$$Av_j = \sum_{i=1}^{j+1} h_{ij} v_i \tag{6}$$

Let V_m be the matrix having n rows and m columns, in which the m columns are the first vectors of the Arnoldi's basis. The orthonormality of vectors, in terms of matrices can be written:

$$^{t}V_m V_m = I_m \tag{7}$$

Likewise, Equation (6), which defines Arnoldi's vectors, can be translated matricially by:

$$AV_m = V_{m+1} H_{m+1.m} \tag{8}$$

Where the matrix $H_{m+1,\, m}$ is a matrix having m+1 rows and m columns where h_{ij} coefficients are the coefficients of orthonormalization of the Equation (6): where $h_{ij} \neq 0$ for i<j+1, and the other coefficients being zero.

$$H_{m+1,m} = \begin{pmatrix} h_{11} & h_{12} & \cdots & \cdots & \cdots & h_{1m} \\ h_{21} & h_{22} & \cdots & \cdots & \cdots & h_{2m} \\ 0 & h_{32} & \ddots & & & \vdots \\ \vdots & \ddots & \ddots & \ddots & & \vdots \\ \vdots & & \ddots & \ddots & \ddots & \vdots \\ 0 & \cdots & \cdots & 0 & h_{mm-1} & h_{mm} \\ 0 & \cdots & \cdots & \cdots & 0 & h_{m+1m} \end{pmatrix}. \tag{9}$$

The main diagonal block of $H_{m+1,\,m}$ is a square matrix of dimension m, noted H_m, which, according the Equations (7) and (8), verifies:

$$H_m = {}^tV_m A V_m = \begin{pmatrix} h_{11} & h_{12} & \cdots & \cdots & \cdots & h_{1m} \\ h_{21} & h_{22} & \cdots & \cdots & \cdots & h_{2m} \\ 0 & h_{32} & \ddots & & & \vdots \\ \vdots & \ddots & \ddots & \ddots & & \vdots \\ \vdots & & \ddots & \ddots & \ddots & \vdots \\ 0 & \cdots & \cdots & 0 & h_{mm-1} & h_{mm} \end{pmatrix}. \tag{10}$$

Such a matrix is called "upper hessenberg": $h_{i,j} = 0 \, \forall i > j+1$.

Equation (10) shows that H_m is the matrix of the projection in Krylov's space K_m of the linear map associated to the matrix A, in the basis $V_m = \{v_1, v_2, , v_m\}$ of Arnoldi's vectors. This can be summed up as follows:

Proposition 1: In Arnoldi's algorithm (A_{Arn}), we have at the step k the following relations:

$$\begin{cases} AV_k = V_k H_k + h_{k+1,k}\, v_{k+1}\, {}^t e_k^{(k)} \\ AV_k = V_{k+1} H_{k+1,k} \\ {}^tV_k A V_k = H_k \end{cases} \tag{11}$$

Where $e_k^{(k)}$ is the vector of the canonical basis of $\mathbb{R}^k$

Proof: [2].

Arnoldi's Method for Linear Systems (FOM)

In this paragraph, we will remind the important results needed for the resolution of linear systems (1).

For that, we will begin by applying the Arnoldi's algorithm (A_{Arn}) to the matrix A, this allows us to obtain at each step k, a matricial pair (V_k,

$H_{k+1,k}$) where V_k is an orthonormal basis of Krylov's space K_{k+1}, and $H_{k+1,k}$ is specific an upper hessenberg matrix given by (9).

The full orthogonalization method (FOM) is variant of problem (3), verifying the space condition (4) and "Petrov-Galerkin condition" (5) by taking $L_m = K_m = K_m(A, r_0)$. This method is defined by:

$$\begin{cases} \text{data} : x_0 \in \mathbb{R}^n \\ \forall k = 1, \cdots, m \leq n \\ \text{find } x_k \in x_0 + K_k \text{ such that :} \\ \qquad r_k \perp K_k \end{cases} \tag{12}$$

Production of the Solution

We are seeking the solution x_m verifying the space condition (4) and "Petrov-Galerkin condition" (5).

(4) $\Leftrightarrow x_m \in x_0 + K_m \Leftrightarrow x_m = x_0 + V_m y$ with $y \in \mathbb{R}^m$.

So, the residual vector $r_m = b - Ax_m$ is written:

$$r_m = r_0 - AV_m y, \tag{13}$$

and "Petrov-Galerkin condition" becomes:

(5) $\Leftrightarrow {}^t V_m r_m = 0 \Leftrightarrow r_m \perp K_m$, with (13)

$\Leftrightarrow {}^t V_m (r_0 - AV_m y) = 0$ but then

$\Leftrightarrow \left({}^t V_m v_1\right)\beta_1 = H_m y$. This is equivalent to:

$$H_m y = \beta_1 e_1^{(m)}. \tag{14}$$

So, (14) implies $y = \beta_1 H_m^{-1} e_1^{(m)}$, and the solution x_m can be written:

$$x_m = x_0 + \beta_1 V_m H_m^{-1} e_1^{(m)} \tag{15}$$

and the problem becomes:

$$\begin{cases} \text{Given } x_0 \in \mathbb{R}^n : \\ \bullet \text{ Seek } y \in^m \text{ verifying } (14) : H_m y = \beta_1 e_1^{(m)}. \\ \bullet \text{ Compute the solution } x_m = x_0 + V_m y. \end{cases}$$

A method based on this approach is called a full orthogonalization method (FOM). It is given by Y.Saad in 1981.

This is the corresponding algorithm using the Arnoldi's process (A_{Arn}).

$$(A_{fom}) : \begin{cases} \text{data} : n,\ A \in \mathbb{R}^{n\times n},\ m \le n,\ V_m \in \mathbb{R}^{n\times m} \\ x_0 \in \mathbb{R}^n,\ r_0 = b - Ax_0,\ h_{jp1j} = 1.0,\ j = 1 \\ \bullet \text{ compute} : v_1 = \dfrac{r_0}{\|r_0\|_2},\ \varepsilon = 10^{-9},\ V_1 = \{v_1\}. \\ \quad \text{while } (h_{jp1j} \ge \varepsilon) \\ \quad\quad \bullet \text{ compute} : w = Av_j \\ \quad\quad \text{for } i = 1 \text{ to } j \\ \quad\quad\quad \bullet \text{ compute} : h_{ij} = \langle w, v_i \rangle \\ \quad\quad\quad \bullet \text{ compute} : w = w - h_{ij} v_i \\ \quad\quad \text{end } i \\ \quad\quad \bullet \text{ compute} : h_{j+1,j} = \|w\|_2 \\ \quad\quad \text{if } h_{j+1,j} = 0. \text{ "Lucky breakdown" then } m = j;\text{ goto } \circledR \\ \quad\quad \bullet \text{ compute} : v_{j+1} = \dfrac{w}{h_{j+1,j}},\ V_{j+1} := \{v_1, \cdots, v_{j+1}\}. \\ \quad\quad h_{jp1j} = h_{j+1,j},\ j = j+1 \\ \quad \text{end while} \\ // \text{ output} : m = j,\ \text{matrices} : V_m, H_m \\ \circledR : \text{compute} : H_m y = \beta_1 e_1 \\ \quad\ \text{compute} : x_m = x_0 + V_m y. \end{cases}$$

The algorithm (A_{fom}) depends on parameter m, which is the dimension of the Krylov's subspace. In practice, it is desirable to select m in a dynamic way. This is possible if the norm of the residue r_m of the solution x_m is calculated with a numerically low cost. (Without having to compute x_m itself). Then the algorithm can be stopped at the appropriate step using this information. The following result gives an answer in this direction.

Proposition 2:
The residue vector r_m of the approximate solution x_m calculated by the algorithm (A_{fom}) is such that:

$$r_m = b - Ax_m = -h_{m+1,m}\, {}^t e_m^{(m)} y v_{m+1},$$

and so,

$$\|b - Ax_m\|_2 = h_{m+1,m} \left| {}^t e_m^{(m)} y \right|. \tag{16}$$

Proof: [2].
At this stage, the interesting feature is the possibility to compute the norm of residue without producing the solution.

According to this proposition, at the step k, the residual norm $\|r_k\|_2$ is then given directly by the absolute value of the last component y_k of the vector $y \in \mathbb{R}^k$. This provides the stopping test and allows to calculate x_k only when the stopping criterion used for the solution is satisfied. This is the main idea of this article. It consists in calculating locally, at each iteration k, uniquely the last component y_k of the vector $y \in \mathbb{R}^k$: solution of the system (14) $H_k y = \beta_1 e_1^{(k)}$. The classical formulae of Cramer's rule directly gives the component y_k in the form of a determinant quotient: $y_k = \frac{\det(D_k)}{\det(H_k)}$, the computation of determinants $\det(D_k)$ is given by the following result:

Proposition 3:

Let $H_k \in \mathbb{R}^{k \times k}$ be an invertible upper hessenberg matrix and $\beta_1 e_1^{(k)} \in \mathbb{R}^k$, the second member of the linear system (14) $H_k y = \beta_1 e_1^{(k)}$. Then we have:

$$y_k = \frac{\begin{vmatrix} h_{1,1} & \cdots & & \cdots & h_{1,k-1} & \beta_1 \\ h_{2,1} & h_{2,2} & & & \vdots & 0 \\ & h_{3,2} & & \ddots & \vdots & \vdots \\ & & & \ddots & h_{k-1,k-1} & \vdots \\ & & & & h_{k,k-1} & 0 \end{vmatrix}}{det(H_k)} = \frac{det(D_k)}{det(H_k)}. \tag{17}$$

with:

$$det(D_k) = (-1)^{k+1} \beta_1 \prod_{i=2}^{k} h_{i,i-1}. \tag{18}$$

$$det(D_{k+1}) = -h_{k+1,k} det(D_k) \quad \forall k \geq 2. \tag{19}$$

What we need now, is to calculate det(H_k), H_k being an upper hessenberg matrix.

STURM SEQUENCE FOR THE UPPER HESSENBERG MATRICES (U.H)

In this paragraph, we extend the definition of Sturm sequence to the hessenberg matrices, and then we give an algorithm computing their determinants. Let's start by reminding the following principal result:

Proposition 4:

Let $H \in \mathbb{R}^{n\times n}$ be an (U.H) matrix and let $(H-\lambda I)_k$ be the kth principal submatrix of $(H-\lambda I)$, i.e.,

$$(H-\lambda I)_k = H_k - \lambda I_k = \begin{pmatrix} h_{1,1}-\lambda & h_{1,2} & \cdots & \cdots & h_{1,k} \\ h_{2,1} & h_{2,2}-\lambda & & & \vdots \\ 0 & h_{3,2} & \ddots & & \vdots \\ \vdots & \ddots & \ddots & \ddots & h_{k-1,k} \\ 0 & \cdots & 0 & h_{k,k-1} & h_{k,k}-\lambda \end{pmatrix}$$

and let $P_k(\lambda)$ be the charcteristic polynomial of H_k, then: the polynomial $P_k(\lambda)$, k=0, ,n are given by to the following recurrent formula:

$$\begin{cases} P_0(\lambda) = 1 \\ P_1(\lambda) = h_{1,1} - \lambda \\ P_k(\lambda) = \sum_{i=1}^{k-1} (-1)^{(k-i)} h_{i,k} \prod_{j=i+1}^{k} h_{j,j-1} P_{i-1}(\lambda) + (h_{k,k}-\lambda) P_{k-1}(\lambda) \end{cases} \tag{20}$$

Proof: by recurrence.

Definition 1:

The sequence $\{P_k(\lambda)\}_{k=0}^{n}$ defined by (20) is called Sturm sequence of the upper hessenberg matrix.

Remark 1:

We note that, for the programming, we can win $\frac{(n-1)(n-2)}{2}$ memory words if we use a storage numbering function [8] for upper hessenberg matrices by storing nonzero elements uniquely.

Computing the Determinant of the Hessenberg Matrix H_n

For calculating the determinant of an upper hessenberg matrix, Proposition (2) gives us a simple method in o (n^2). This can be summed by the following result:

Corollary 1:

Let $H = (h_{i,j}) \in \mathbb{R}^{n \times n}$ be an invertible (U.H) matrix. If we set $\delta_k = \det(H_k)$, then we have:

$$\begin{cases} \delta_0 = 1 \\ \delta_1 = h_{11} \\ \delta_k = h_{kk}\delta_{k-1} + \sum_{\ell=2}^{k} (-1)^{(\ell+1)} h_{k+1-\ell,k}\delta_{k-\ell} \prod_{j=k+2-\ell}^{k} h_{j,j-1} \quad \forall k = 2,\cdots,n. \end{cases} \tag{21}$$

$\delta_n = \det(H)$. In the Formula (21), we calculate the products $\prod_{j=k+2-\ell}^{k} h_{j,j-1}$, without repeting multiplications. So, we obtain a very interesting algorithm (A_{det}) computing $\delta_n = \det(H_n)$ as follows:

$$(A_{\det}):\begin{cases}\text{Data}: n,\ H_n \in \mathbb{R}^{n\times n}\ (\text{U.H}),\ \ \delta \in \mathbb{R}^{n+1} \\ \qquad \delta_0 = 1.0,\ \delta_1 = h_{1.1} \\ \text{for}\ k = 1\ \text{ to }\ n-1\ \text{ do}: \\ \quad \text{begin} \\ \qquad S = h_{kk} * \delta_k,\ \ codiag = 1.0,\ \ sg = -1 \\ \qquad \text{for}\ i = k-1\ \text{step}\ -1,\ 0\ \ \text{do}: \\ \qquad\quad \text{begin} \\ \qquad\qquad codiag = codiag * h_{i+1.i} \\ \qquad\qquad P = h_{i.k} * codiag * \delta_i \\ \qquad\qquad S = S + sg * P,\ sg = -sg \\ \qquad\quad \text{end}\{i\} \\ \qquad \delta_{k+1} = S \\ \quad \text{end}\{k\} \\ //\,\text{Output}:\ \ det(H_n) = \delta_n \end{cases}$$

The number of operations necessary in calculating δ_n is:

1. Additions: $\sum_{k=2}^{n}(k-1) = \frac{n(n-1)}{}$
2. Multiplications: $\sum_{k=1}^{n} 5k = \frac{5\hat{n}(n+1)}{2}$.

The total number of operations required is $\sim 3n^2$. This is more convenient when n is big.

We will adopt the algorithm (A_{det}) in the full orthogonalization method (FOM). Having at each step k, $(V_{k+1}, H_{k+1,k})$, we calculate by the Formulaes (18) and (19): $y_k = \frac{\det(D_k)}{\det(H_k)}$. This gives us the stopping test and allows to calculate x_k only when the stopping criterion used for the solution is satisfied, namely:

$$\|r_k\|_2 = h_{k+1,k} * |y_k| < \varepsilon$$

Where ε, being the required precision. The FOM algorithm (A_{Fom}) takes the following form:

$$
(A_{Fom}):\left\{\begin{array}{l}
\text{Data}: n,\ A\in^{n\times n},\ m\le n,\ \varepsilon=10^{-8}\ (\text{for example})\\
x_0\in^{n},\ \delta\in^{n+1}\ H\in^{m+1\times m},\ V\in^{n\times m},\ sg\in\{-1,1\}: sign\\
\text{Initialization}: r_0=b-Ax_0,\ \ \delta_0=1.0,\\
\text{compute}: \beta_1=\|r_0\|_2,\ \ v_0=\dfrac{r_0}{\beta_1},\ \ V_0=\{v_0\}.\\
//\text{calculate the column } h_0 \text{ of the matrix } H_0\\
\quad \text{compute}: w=Av_0\ ,\ \ h_{00}=\langle w,v_0\rangle.\\
\quad \text{compute}: w=w-h_{00}v_0\\
\quad \text{compute}: h_{1,0}=\|w\|_2\ \ \text{if } h_{1,0}=0,\text{"Luckybreakdown"},\text{goto} ®\\
\quad \text{compute}: v_1=\dfrac{w}{h_{1,0}},\ \ \ H=[h_0],V_1=[v_0,v_1]\\
\quad k=1;\ \delta_1=h_{0,0},\ detDkm1=\beta_1,\ \|r_k\|_2=h_{0,0};\\
\quad \text{while}\left(\|r_k\|_2>\varepsilon\right)\\
\quad //\text{calculate the colomn } h_k \text{ of the matrix } H_k\\
\quad \text{compute}: w=Av_k\\
\quad \text{for } i=0\ \text{ to }\ k\\
\quad \text{compute}: h_{ik}=\langle w,v_i\rangle\\
\quad \text{compute}: w=w-h_{ik}v_i\\
\quad \text{end } i\\
\quad \text{compute}: h_{k+1,k}=\|w\|_2,\ \text{if } h_{k+1,k}=0,\text{"Luckybreakdown"},\ \text{goto} ®\\
\quad \text{compute}: v_{k+1}=\dfrac{w}{h_{k+1,k}},\ \ \begin{cases}V_{k+1}=[v_0,\cdots,v_{k+1}]\\ H_k=[h_0,\cdots,h_k]\end{cases}\\
\quad //\text{compute } \det(H_k)\\
\quad detHk=h_{k,k}*\delta_k;\ \ codiag=1.0;\ \ sg=-1;\\
\quad \text{for } i=k-1\ \text{ to }\ 0\ \ \text{step} -1\\
\quad codiag=codiag*h_{i+1,i};\\
\quad P=h_{ik}*codiag*\delta_i;\\
\quad detHk=detHk+sg*P;\ \ sg=-sg;\\
\quad \text{end } i\\
\quad \delta_{k+1}=detHk;\ \ \ detDk=-detDkm1*h_{k,k-1};\\
\quad y_k=detDk/detHk;\ \ \|r_k\|_2=h_{k+1,k}*|y_k|;\\
\quad detDkm1=detDk,\ \ k:=k+1;\\
\quad \text{end}\{\text{while}\}\\
®\ \text{Output}: m=k,\ \ \text{resolve}: H_m y=\beta_1 e_1^{(m)}.
\end{array}\right.
$$

When the stopping criterion used for the solution is satisfied, we take the value y_k for calculating $y_{k-1}, y_{k-2}, \cdots, y_1$, by using the following algorithm:

Algorithm of Resolution of the System: $H_m y = \beta_1 e_1$

$$
(A_{Res}): \begin{cases}
\text{Data}: \begin{cases} n \geq m \geq 2; \quad \beta_1 \in \mathbb{R} \\ V_m = [v_0, \cdots, v_{m-1}] \in M_{n.m}(\mathbb{R}) \\ H_m = [h_0, \cdots, h_{m-1}] \in M_m(\mathbb{R}) \\ y \in \mathbb{R}^m,\ y_m \in \mathbb{R}, \text{ calculated already} \\ x_0 \in \mathbb{R}^n, \quad x \in \mathbb{R}^n, \quad b = \beta_1 e_1^{(m)} \in \mathbb{R}^m. \end{cases} \\
\bullet \text{ compute the system solution}: Hy = b \\
\forall\, i = m-1, \cdots, 1, \text{ step } -1 \\
\qquad y_{i-1} = -\dfrac{\sum_{j=i}^{m-1} h_{ij} y_j}{h_{i,i-1}} \\
A_i \\
// \text{ being } y \in \mathbb{R}^m, \text{ compute the solution } x \in \mathbb{R}^n \\
\bullet \text{ compute}: x = x_0 + V_m * y;
\end{cases}
$$

The above algorithm (A_{Res}) requires 2nm flops.

Several authors solve the linear system $H_m y = \beta_1 e_1^{(m)}$ using Givens method applied to an upper hessenberg matrix.

THE GIVENS METHOD FOR (U.H) MATRICES

In this section, we give a new variant of the Givens method for the resolution of linear systems with an upper hessenberg matrices coming from Arnoldi's algorithm. This method is based on the addition of two matrices and two tensorial products.

The Givens rotations constitute an important tool to selectively eliminate certain coefficients of a matrix. This is the case of our upper hessenberg matrix, where we eliminate the codiagonal in order to obtain an upper triangular matrix.

To justify the next results, we introduce the following notations:

- $\tilde{\alpha}_k^{(r)}$: The kth row of the matrix $A^{(r)}$ at the step r.
- $\tilde{\tilde{\alpha}}^{(r)}$: The matrix $A^{(r)}$ with the rth and (r+1)^th^ row put to zero.
- α_k : k^th^ column vector of A.
- ${}^tX$: the superscript t will denote the matrix or the transposed vector.
- $x \otimes^t y$: Tensor product of two vectors x, y in $\mathbb{R}^n$.
- $[g_{ij}]_k$: k^th^ column vector of the rotation matrix $G_{i,j}$ in the plan $\{e_i, e_j\}$.

Givens Formulation for Upper Hessenberg Matrices

The classical formulation of the Givens method for an upper hessenberg matrix $H \in^{n\times n}$ is given under the following form:

$$\begin{cases} H^{(1)} = H \\ H^{(2)} = G_{21} \times H^{(1)} \\ H^{(3)} = G_{32} \times H^{(2)} \\ \quad\vdots \\ H^{(k+1)} = G_{k+1,k} \times H^{(k)} \\ \quad\vdots \\ H^{(n-1)} = G_{n-1,n-2} \times H^{n-2} \\ H^{(n)} = G_{n,n-1} \times H^{(n-1)} \end{cases} \tag{22}$$

with $G_{k+1,\,k}$ is of the form:

$$G_{k+1,k} = \begin{array}{r} \\ \\ \\ k \rightarrow \\ k+1 \rightarrow \\ \\ \\ \\ \end{array} \begin{pmatrix} 1 & & & \overset{\substack{k \\ \downarrow}}{\vdots} & \overset{\substack{k+1 \\ \downarrow}}{\vdots} & & & \\ & \ddots & & \vdots & \vdots & & O & \\ & & 1 & \vdots & \vdots & & & \\ \cdots & \cdots & \cdots & c_k & s_k & \cdots & \cdots & \cdots \\ \cdots & \cdots & \cdots & -s_k & c_k & \cdots & \cdots & \cdots \\ & & & \vdots & \vdots & 1 & & \\ & O & & \vdots & \vdots & & \ddots & \\ & & & \vdots & \vdots & & & 1 \end{pmatrix} \tag{23}$$

with $\det(G_{k+1,\,k})=1$ and where c_k and s_k are calculated at each step k by the following formulas:

$$\begin{cases} d_{k+1,k}^{(k)} = \sqrt{\left(h_{k,k}^{(k)}\right)^2 + \left(h_{k+1,k}^{(k)}\right)^2}. \\ c_k = \dfrac{h_{k,k}^{(k)}}{d_{k+1,k}^{(k)}}. \\ s_k = \dfrac{h_{k+1,k}^{(k)}}{d_{k+1,k}^{(k)}}. \end{cases} \tag{24}$$

The relations (22) give us an upper triangular matrix, $H^{(n)}$:

$$H^{(n)} = G_{n,n-1}G_{n-1,n-2}\cdots G_{3,2}G_{2,1}H, \tag{25}$$

and therefore

$$H = G_{21}^{-1}G_{32}^{-1}\cdots G_{n-1,n-2}^{-1}G_{n,n-1}^{-1}H^{(n)}.$$

As the matrices G_{ij} are orthogonal, we have: ${}^{t}G_{ij}G_{ij} = I_{d'}$, from which

$$H = {}^tG_{21}\,{}^tG_{32}\cdots{}^tG_{n-1,n-2}\,{}^tG_{n,n-1}H^{(n)}. \tag{26}$$

Theorem 1:

If $H \in \mathbb{R}^{n,n}$ is an invertible upper hessenberg matrix, then $\forall 1 \le k \le n-1$

$$H^{(k+1)} = G_{k+1,k}H^{(k)} \Leftrightarrow H^{(k+1)} = \tilde{H}^{(k)} + \left[g_{k+1,k}\right]_k \otimes \tilde{h}_k^{(k)} + \left[g_{k+1,k}\right]_{k+1} \otimes \tilde{h}_{k+1}^{(k)}. \tag{27}$$

Where $\tilde{h}_k^{(k)}$ is the kth row vector of the matrix $H^{(k)}$ and $[g_{k+1,k}]_k$ is the kth column vector of the matrix $G_{k+1,k}$.

The Givens algorithm for the triangularisation of the matrix $H \in \mathbb{R}^{n,n}$ (U.H), is simplified to be written in its new form:

$$(A_{Giv}):\left\{\begin{array}{l}
\bullet\text{Data}: n,\ H \in^{n\times n+1} \text{ augmented (U.H) matrix of } b\\
\text{for } k=1 \text{ to } n-1 \text{ do}:\\
\quad \text{If } h_{k+1,k}^{(k)} = 0.0 \quad \text{then : continue}\\
\quad \bullet \text{ calculation of the rotation matrix (23) } G_{k+1,k}^{(k)}\\
\quad\quad G_{k+1,k}^{(k)}:\left\{\begin{array}{l}
\bullet \text{ compute } d_{k+1,k}^{(k)} = \sqrt{\left(h_{k,k}^{(k)}\right)^2 + \left(h_{k+1,k}^{(k)}\right)^2}\\
\bullet \text{ compute } c_k = \dfrac{h_{k,k}^{(k)}}{d_{k+1,k}^{(k)}},\ s_k = \dfrac{h_{k+1,k}^{(k)}}{d_{k+1,k}^{(k)}}
\end{array}\right.\\
\quad \bullet \text{ compute}: \left[g_{k+1,k}\right]_k = \begin{pmatrix} 0\\ \vdots\\ 0\\ c_k\\ -s_k\\ 0\\ \vdots\\ 0\end{pmatrix} \leftarrow k : \left[g_{k+1,k}\right]_{k+1} = \begin{pmatrix} 0\\ \vdots\\ 0\\ c_k\\ s_k\\ 0\\ \vdots\\ 0\end{pmatrix} \leftarrow k\\
\quad \bullet\text{compute}: H^{(k+1)} = \tilde{H}^{(k)} + \left[g_{k+1,k}\right]_k \otimes \tilde{h}_k^{(k)} + \left[g_{k+1,k}\right]_{k+1} \otimes \tilde{h}_{k+1}^{(k)}\\
fin\{k\}
\end{array}\right.$$

Complexity of the Algorithm (A_{Giv})

To eliminate the (n – 1) elements situated under the diagonal, we need (n – 1) rotations $G^{(k)}_{k+1,k}$. Each rotation requires 5 operations and an extraction of a square root $(\sqrt{\ })$.

For a fixed k

→ Computation of tensor product:

$$\left[g_{k+1,k}\right]_k \otimes \tilde{h}_k^{(k)}: \ 2(n-k+2)(*)$$

$$\left[g_{k+1,k}\right]_{k+1} \otimes \tilde{h}_{k+1}^{(k)}: \ 2(n-k+2)(*)$$

→ Computation of matricial addition:

$$\left[g_{k+1,k}\right]_k \otimes \tilde{h}_k^{(k)} + \left[g_{k+1,k}\right]_{k+1} \otimes \tilde{h}_{k+1}^{(k)}: \ 2(n-k+2)(+)$$

and $\forall k = 1,\cdots,n-1$, we have:

$$6\sum_{k=1}^{n-1}(n-k+2)+\sum_{k=1}^{n-1}5 \simeq 6\int_{k=0}^{n-2}(n-k+2)\,dk \simeq 3n^2$$

Conclusion: The complexity of Givens' algorithm (A_{Giv}) for an upper hessenberg matrix requires $3n^2$ flops and $(n-1)(\sqrt{\ })$.

In consequense, we find the calculating formula of the determinant of the matrix H given by:

Corollary 2:

Let H be an (U.H) matrix. By applying classical Givens' formulation, given by the relation (22) and the Formula (26), we obtain:

$$\det(H) = \det\left(H^{(n)}\right) = \prod_{i=1}^{n} h_{i,i}^{(i)}. \quad (28)$$

We can decompose H matrix (U.H) of rank n, into an addition of a triangular matrix and (n-1) matrices of rang 1 given by:

$$H = H^{(n+1)} - \sum_{k=1}^{n-1} \left(\left[g_{k+1,k} \right]_k - e_k \right) \otimes \tilde{h}_k^{(k)} + \left(\left[g_{k+1,k} \right]_{k+1} - e_{k+1} \right) \otimes \tilde{h}_{k+1}^{(k)}. \tag{29}$$

Numerical Example

Let $H \in \mathbb{R}^{4\times4}$ be an invertible matrix (U.H), and $b \in \mathbb{R}^4$.

We want to resolve Hx=b.

$$H = \begin{pmatrix} 1 & 0 & -1 & 2 \\ 1 & 2 & -3 & 0 \\ 0 & 1 & -1 & 0 \\ 0 & 0 & -1 & 1 \end{pmatrix}; \quad b = \begin{pmatrix} 6 \\ -4 \\ -1 \\ 1 \end{pmatrix}$$

We apply the algorithm (A_{Giv}) to the augmented matrix $H \in \mathbb{R}^{4\times5}$ by the second member b.

Step k = 1

– Compute the rotation in plane $\{e_2, e_1\}$: $G_{21} = \begin{pmatrix} \frac{\sqrt{2}}{2} & \frac{\sqrt{2}}{2} \\ -\frac{\sqrt{2}}{2} & \frac{\sqrt{2}}{2} \end{pmatrix}$

Column 1: $[g_{21}]_1 = \begin{pmatrix} \frac{\sqrt{2}}{2} \\ -\frac{\sqrt{2}}{2} \end{pmatrix}$; column 2: $[g_{21}]_2 = \begin{pmatrix} \frac{\sqrt{2}}{2} \\ \frac{\sqrt{2}}{2} \end{pmatrix}$

$$\underbrace{\begin{pmatrix} 0 & 0 & 0 & 0 & 0 \\ 0 & 0 & 0 & 0 & 0 \\ 0 & 1 & -1 & 0 & -1 \\ 0 & 0 & -1 & 1 & 1 \end{pmatrix}}_{\tilde{H}^{(1)}} + \begin{array}{c|ccccc} \tilde{h}_1^{(1)}: & 1 & 0 & -1 & 2 & 6 \\ \hline \frac{\sqrt{2}}{2} & \frac{\sqrt{2}}{2} & 0 & -\frac{\sqrt{2}}{2} & 2\frac{\sqrt{2}}{2} & 6\frac{\sqrt{2}}{2} \\ -\frac{\sqrt{2}}{2} & -\frac{\sqrt{2}}{2} & 0 & \frac{\sqrt{2}}{2} & -2\frac{\sqrt{2}}{2} & -6\frac{\sqrt{2}}{2} \\ 0 & 0 & 0 & 0 & 0 & 0 \\ 0 & 0 & 0 & 0 & 0 & 0 \\ [g_{21}]_1 & & & [g_{21}]_1 \otimes \tilde{h}_1^{(1)} & & \end{array} + \begin{array}{c|ccccc} \tilde{h}_2^{(1)}: & 1 & 2 & -3 & 0 & -4 \\ \hline \frac{\sqrt{2}}{2} & \frac{\sqrt{2}}{2} & \sqrt{2} & -3\frac{\sqrt{2}}{2} & 0 & -4\frac{\sqrt{2}}{2} \\ \frac{\sqrt{2}}{2} & \frac{\sqrt{2}}{2} & \sqrt{2} & -3\frac{\sqrt{2}}{2} & 0 & -4\frac{\sqrt{2}}{2} \\ 0 & 0 & 0 & 0 & 0 & 0 \\ 0 & 0 & 0 & 0 & 0 & 0 \\ [g_{21}]_2 & & & [g_{21}]_2 \otimes \tilde{h}_2^{(1)} & & \end{array}$$

$$
= \underbrace{\begin{pmatrix} \sqrt{2} & \sqrt{2} & -2\sqrt{2} & \sqrt{2} & \sqrt{2} \\ 0 & \sqrt{2} & -\sqrt{2} & -\sqrt{2} & -5\sqrt{2} \\ 0 & 1 & -1 & 0 & -1 \\ 0 & 0 & -1 & 1 & 1 \end{pmatrix}}_{H^{(2)}}
$$

Step k = 2

Compute the rotation in plane $\{e_3, e_2\}$: $G_{32} = \begin{pmatrix} \frac{\sqrt{2}}{\sqrt{3}} & \frac{1}{\sqrt{3}} \\ \frac{-1}{\sqrt{3}} & \frac{\sqrt{2}}{\sqrt{3}} \end{pmatrix}$

Column 2: $[g_{32}]_2 = \begin{pmatrix} \frac{\sqrt{2}}{\sqrt{3}} \\ \frac{-1}{\sqrt{3}} \end{pmatrix}$; column 3: $[g_{32}]_3 = \begin{pmatrix} \frac{1}{\sqrt{3}} \\ \frac{\sqrt{2}}{\sqrt{3}} \end{pmatrix}$

$$
\underbrace{\begin{pmatrix} \sqrt{2} & \sqrt{2} & -2\sqrt{2} & \sqrt{2} & \sqrt{2} \\ 0 & 0 & 0 & 0 & 0 \\ 0 & 0 & 0 & 0 & 0 \\ 0 & 0 & -1 & 1 & 1 \end{pmatrix}}_{\tilde{H}^{(2)}}
+
\begin{array}{c|ccccc}
\tilde{h}_2^{(2)}: & 0 & \sqrt{2} & -\sqrt{2} & -\sqrt{2} & -5\sqrt{2} \\ \hline
0 & 0 & 0 & 0 & 0 & 0 \\
\frac{\sqrt{2}}{\sqrt{3}} & 0 & \frac{2}{\sqrt{3}} & \frac{-2}{\sqrt{3}} & \frac{-2}{\sqrt{3}} & \frac{-10}{\sqrt{3}} \\
\frac{-1}{\sqrt{3}} & 0 & \frac{-\sqrt{2}}{\sqrt{3}} & \frac{\sqrt{2}}{\sqrt{3}} & \frac{\sqrt{2}}{\sqrt{3}} & \frac{5\sqrt{2}}{\sqrt{3}} \\
0 & 0 & 0 & 0 & 0 & 0 \\
{[g_{32}]_2} & \multicolumn{5}{c}{[g_{32}]_2 \otimes \tilde{h}_2^{(2)}}
\end{array}
$$

$$
+
\begin{array}{c|ccccc}
\tilde{h}_3^{(2)}: & 0 & 1 & -1 & 0 & -1 \\ \hline
0 & 0 & 0 & 0 & 0 & 0 \\
\frac{1}{\sqrt{3}} & 0 & \frac{1}{\sqrt{3}} & \frac{-1}{\sqrt{3}} & 0 & \frac{-1}{\sqrt{3}} \\
\frac{\sqrt{2}}{\sqrt{3}} & 0 & \frac{\sqrt{2}}{\sqrt{3}} & \frac{-\sqrt{2}}{\sqrt{3}} & 0 & \frac{-\sqrt{2}}{\sqrt{3}} \\
0 & 0 & 0 & 0 & 0 & 0 \\
{[g_{32}]_3} & \multicolumn{5}{c}{[g_{32}]_3 \otimes \tilde{h}_3^{(2)}}
\end{array}
=
\underbrace{\begin{pmatrix} \sqrt{2} & \sqrt{2} & -2\sqrt{2} & \sqrt{2} & \sqrt{2} \\ 0 & \frac{3}{\sqrt{3}} & \frac{-3}{\sqrt{3}} & \frac{-2}{\sqrt{3}} & \frac{-11}{\sqrt{3}} \\ 0 & 0 & 0 & \frac{\sqrt{2}}{\sqrt{3}} & \frac{4\sqrt{2}}{\sqrt{3}} \\ 0 & 0 & -1 & 1 & 1 \end{pmatrix}}_{H^{(3)}}
$$

Step k = 3

Compute the rotation in plane $\{e_4, e_3\}$: $G_{43} = \begin{pmatrix} 0 & 1 \\ -1 & 0 \end{pmatrix}$

Column 3: $[g_{43}]_3 = \begin{pmatrix} 0 \\ -1 \end{pmatrix}$; column 4: $[g_{43}]_4 = \begin{pmatrix} 0 \\ 1 \end{pmatrix}$

$$\underbrace{\begin{pmatrix} \sqrt{2} & \sqrt{2} & -2\sqrt{2} & \sqrt{2} & \sqrt{2} \\ 0 & \frac{3}{\sqrt{3}} & \frac{-3}{\sqrt{3}} & \frac{-2}{\sqrt{3}} & \frac{-11}{\sqrt{3}} \\ 0 & 0 & 0 & 0 & 0 \\ 0 & 0 & 0 & 0 & 0 \end{pmatrix}}_{\tilde{H}^{(3)}} + \begin{array}{c|ccccc} \tilde{h}_3^{(3)}: & 0 & 0 & 0 & \frac{\sqrt{2}}{\sqrt{3}} & \frac{4\sqrt{2}}{\sqrt{3}} \\ \hline 0 & 0 & 0 & 0 & 0 & 0 \\ 0 & 0 & 0 & 0 & 0 & 0 \\ 0 & 0 & 0 & 0 & 0 & 0 \\ -1 & 0 & 0 & 0 & \frac{-\sqrt{2}}{\sqrt{3}} & \frac{-4\sqrt{2}}{\sqrt{3}} \\ [g_{43}]_3 & & & [g_{43}]_3 \otimes \tilde{h}_3^{(3)} & & \end{array}$$

$$+ \begin{array}{c|ccccc} \tilde{h}_4^{(3)}: & 0 & 0 & -1 & 1 & 1 \\ \hline 0 & 0 & 0 & 0 & 0 & 0 \\ 0 & 0 & 0 & 0 & 0 & 0 \\ 1 & 0 & 0 & -1 & 1 & 1 \\ 0 & 0 & 0 & 0 & 0 & 0 \\ [g_{43}]_4 & & [g_{43}]_4 \otimes \tilde{h}_4^{(3)} & & & \end{array} = \underbrace{\begin{pmatrix} \sqrt{2} & \sqrt{2} & -2\sqrt{2} & \sqrt{2} & \sqrt{2} \\ 0 & \frac{3}{\sqrt{3}} & \frac{-3}{\sqrt{3}} & \frac{-2}{\sqrt{3}} & \frac{-11}{\sqrt{3}} \\ 0 & 0 & -1 & 1 & 1 \\ 0 & 0 & 0 & -\frac{\sqrt{2}}{\sqrt{3}} & \frac{-4\sqrt{2}}{\sqrt{3}} \end{pmatrix}}_{H^{(4)}}$$

We obtain a triangular system $H^{(4)}x=b^{(4)}$ whose solution is x:

$$H^{(4)} = \begin{pmatrix} \sqrt{2} & \sqrt{2} & -2\sqrt{2} & \sqrt{2} \\ 0 & \frac{3}{\sqrt{3}} & \frac{-3}{\sqrt{3}} & \frac{-2}{\sqrt{3}} \\ 0 & 0 & -1 & 1 \\ 0 & 0 & 0 & \frac{-\sqrt{2}}{\sqrt{3}} \end{pmatrix}; \quad b^{(4)} = \begin{pmatrix} \sqrt{2} \\ \frac{-11}{\sqrt{3}} \\ 1 \\ \frac{-4\sqrt{2}}{\sqrt{3}} \end{pmatrix}; \quad x = \begin{pmatrix} 1 \\ 2 \\ 3 \\ 4 \end{pmatrix}$$

and we have $\det(H) = \det(H^{(4)}) = 2$.

NUMERICAL TESTS

In this section, we present four numerical tests for the resolution of linear systems (1): AX=b for the pro–posed new version of full orthogonalization method (FOM). These tests are done using a portable DELL (Latitude D505 intel (R) Pentium (R) M, processor 1.70 GHz, 594 MHz) and a program written in C_{++}, with double precision.

Test 1: A_n is band (SPD) matrix D=2d+1, d=3, [9] and $X \in \mathbb{R}^n$ is the exact solution.

$$A_n = \begin{pmatrix} 5 & 2 & 1 & 1 & 0 & \cdots & 0 \\ 2 & 6 & 3 & 1 & 1 & \ddots & \vdots \\ 1 & 3 & 6 & \ddots & \ddots & \ddots & 0 \\ 1 & \ddots & \ddots & \ddots & \ddots & \ddots & 1 \\ 0 & \ddots & \ddots & \ddots & \ddots & 3 & 1 \\ \vdots & \ddots & 1 & 1 & 3 & 6 & 2 \\ 0 & \cdots & 0 & 1 & 1 & 2 & 5 \end{pmatrix}, \quad b = \begin{pmatrix} 16 \\ 2\times16 \\ \vdots \\ (n-3)16 \\ 15n-23 \\ 13n-18 \\ 9n-7 \end{pmatrix}, \quad X = \begin{pmatrix} 1 \\ 2 \\ 3 \\ \vdots \\ n-2 \\ n-1 \\ n \end{pmatrix}$$

From test 1, we deduce the following numerical results in Table 1.

Test 2: A_n is a full matrix, X is the exact solution, and A_n^{-1} is the inverse matrix.

$$A_n = \begin{pmatrix} 1 & \cdots & \cdots & 1 \\ 2 & \ddots & \ddots & \vdots \\ \vdots & \ddots & \ddots & 1 \\ 2 & \cdots & 2 & 1 \end{pmatrix}, \quad X = \begin{pmatrix} 1 \\ 2 \\ \vdots \\ n-1 \\ n \end{pmatrix}, \quad A_n^{-1} = \begin{pmatrix} -1 & 1 & 0 & \cdots & 0 \\ 0 & \ddots & \ddots & \ddots & \vdots \\ \vdots & \ddots & \ddots & \ddots & 0 \\ 0 & \cdots & 0 & \ddots & 1 \\ 2 & 0 & \cdots & 0 & -1 \end{pmatrix}$$

Where: $\|A_n\|_\infty = 2n-1, \|A_n^{-1}\|_\infty = 3$ which imply: $\text{cond}_\infty(A_n) \simeq 6n$.

$$\begin{cases} b_i = \dfrac{2i(i-1)+(n+i)(n-i+1)}{2} \\ \forall i = 1,\cdots,n \end{cases} \tag{30}$$

From test 2, we deduce the following numerical results in Table 2.

Test 3: In the third test, we focus our attention on the resolution of the partial differential equations with Neuman condition:

$$\begin{cases} -\Delta u + u = f & \text{on } \Omega =]0,\gamma[\times]0,\gamma[\\ \dfrac{\partial u}{\partial \nu} = 0 & \text{on } \Gamma \text{ the frontier} \end{cases} \tag{31}$$

$$\text{with:} \begin{cases} h_x = h_y = 0.01 : \text{step of discretization at x and at y.} \\ n \in \mathbb{N}^* : \text{number of nodes on the x axis.} \\ \gamma = h_x(n-1). \end{cases}$$

Table 1: Test 1's numerical results

n	m	ε	$\|r_k\|_2$
100	66	0.910^{-3}	0.810^{-3}
500	171	0.910^{-3}	0.89310^{-3}
1000	215	0.910^{-3}	0.89410^{-3}
1500	247	0.910^{-3}	0.8810^{-3}
2000	270	0.910^{-3}	0.89510^{-3}
2500	291	0.910^{-3}	0.89610^{-3}
3000	309	0.910^{-3}	0.89810^{-3}

Table 2: Test 2's numerical results

n	m	ε	$\|r_k\|_2$	$cond_\infty(A_n) \simeq 6n$
100	47	0.910^{-3}	0.54310^{-3}	600
200	115	0.910^{-3}	0.84710^{-3}	3000
1000	169	0.910^{-3}	0.78810^{-3}	6000
1500	211	0.910^{-3}	0.78510^{-3}	9000
2000	246	0.910^{-3}	0.8710^{-3}	12,000
2500	278	0.910^{-3}	0.8410^{-3}	15,000
3000	307	0.910^{-3}	0.8310^{-3}	18,000

If we consider $u(x,y)=\cos\left(\frac{\pi}{\gamma}x\right)\cos\left(\frac{3\pi}{\gamma}y\right)$ as a solution, we obtain the function:

$$f(x,y)=\left(\frac{10\pi^2+\gamma^2}{\gamma^2}\right)\cos\left(\frac{\pi}{\gamma}x\right)\cos\left(\frac{3\pi}{\gamma}y\right) \tag{32}$$

as a second member.

After doing a regular maillage and numbering the nodes, the calculation of the stiffness matrix leads to a (SPD) "well-conditioned" matrix. For the resolution we use the new version of full orthogonalization method (FOM).

Table 3 contains the numerical results of Test 3:

Test 4: A_n is band, (SPD) matrix D=2d+1, d=2, [9].

$$A_n=\begin{pmatrix} 5 & -4 & 1 & 0 & \cdots & 0 \\ -4 & 6 & \ddots & \ddots & \ddots & \vdots \\ 1 & \ddots & \ddots & \ddots & \ddots & 0 \\ 0 & \ddots & \ddots & \ddots & \ddots & 1 \\ \vdots & \ddots & \ddots & \ddots & 6 & -4 \\ 0 & \cdots & 0 & 1 & -4 & 5 \end{pmatrix},\ b=\begin{pmatrix} 0 \\ \vdots \\ 0 \\ -(n+1) \\ 2(n+1) \end{pmatrix},\ X=\begin{pmatrix} 1 \\ 2 \\ \vdots \\ n-1 \\ n \end{pmatrix}$$

$$\lambda_k=16\sin^4\left(k\pi 2(n+1)\right) \qquad \forall k=1,\cdots,n \tag{33}$$

On the one hand, the numerical results for the matrices A_n are considerably polluted with rounding errors because the matrices A_n are very "ill-conditioned" But on the other hand, we are constructing matrices M_n very "well-conditioned" from matrices A_n with translations:

$$M_n=A_n-\mu_n I_d$$

Where $\mu_n=16\sin^4(n\pi 2(n+1))$.

From Test 4, we deduce the following numerical results in Table 4.

Table 3: Test 3's numerical results

n	m	ε	$\|r_k\|_2$
100	25	0.910^{-3}	0.52410^{-3}
900	75	0.910^{-3}	0.48110^{-3}
1600	94	0.910^{-3}	0.66210^{-3}
2500	114	0.910^{-3}	0.610^{-3}
3600	131	0.910^{-3}	0.8510^{-3}

Table 4: Test 4's numerical results

n	m	ε	$\|r_k\|_2$	$cond_2(A_n)=\frac{\lambda_n}{\lambda_1}$	$cond_2\left(\frac{A_n-\mu_n I_d}{M_n}\right)$
100	10	0.910^{-3}	0.8810^{-3}	1.777810^{7}	$cond_2(M_n)\simeq 2$
500	11	0.910^{-3}	0.7610^{-3}	1.038910^{10}	$cond_2(M_n)\simeq 2$
1000	12	0.910^{-3}	0.2610^{-3}	1.649110^{11}	$cond_2(M_n)\simeq 2$
1500	11	0.910^{-3}	0.39110^{-3}	8.337710^{11}	$cond_2(M_n)\simeq 2$
2000	12	0.910^{-3}	0.5210^{-3}	2.633710^{12}	$cond_2(M_n)\simeq 2$
2500	12	0.910^{-3}	0.6510^{-3}	6.666710^{12}	$cond_2(M_n)\simeq 2$
3000	12	0.910^{-3}	0.7810^{-3}	1.333310^{13}	$cond_2(M_n)\simeq 2$
3500	13	0.910^{-3}	0.15610^{-3}	2.468010^{13}	$cond_2(M_n)\simeq 2$

CONCLUSIONS

Test 1 and 2 show the results of applying the proposed new version of full orthogonalization method (FOM): the dimension m of Krylov's space K_m is acceptable compared to n because the matrices A_n are moderately "ill-conditioned".

For the third test, the stiffness matrix coming from the discretization of the partial differential Equations (31) leads to a (SPD) well-conditioned matrix. We note that dimension m of Krylov's space K_m is weak com–pared to n, which can give a positive judgment about the new version of (FOM) method.

In the final test, the constructed matrices M_n are very "well-conditioned", and the dimension m is as small as required.

ACKNOWLEDGEMENTS

The author is grateful to the referees for their valuable comments and suggestions which have helped to improve the presentation of this work.

REFERENCES

1. Toumi, A. (2005) Utilisation des filtres de Tchebycheff et construction de préconditionneurs spéciaux pour l'accélération des methods de Krylov. Thèse No. 2296, de l'Institut National Polytechnique de Toulouse, France.
2. Saad, Y. (2000) Iterative Methods for Sparse Linear Systems. 2nd Edition, Society for Industrial and Applied Mathematics, Philadelphia.
3. Benhamadou, M. (2000) Développement d'outils en Programmation Linéaire et Analyse Numérique matricielle. Thèse No. 1955, de l'Université Paul Sabatier Toulouse 3, Toulouse, France.
4. Wilkinson, J.H. (1965) the Algebraic Eigenvalue Problem. Clarendon Press, Oxford.
5. Gastinel, N. (1966) Analyse Numérique Linéaire. Hermann, Paris.
6. Ciarlet, P.G. (1980) Introduction à l'Analyse Numérique Matricielle et à l'Optimisation. Masson, Paris.
7. Lascaux, P. and Théodor, R. (1993) Analyse Numérique Matricielle Appliquée à l'Art de l'Ingénieur. Tome 1, Tome 2, Masson, Paris.
8. Jennings, A. (1980) Matrix Computation for Engineers and Scientists. John Wiley and Sons, Chichester.
9. Gregory, R.T. and Karney, D.L. (1969) A Collection of Matrices for Testing Computational Algorithms. Wiley-Inter–science, John Wiley & Sons, New York, London, Sydney, Toronto.

CITATION

1. Benhamadou, M. (2014) On the FOM Algorithm for the Resolution of the Linear Systems Ax = b. Advances in Linear Algebra & Matrix Theory, 4, 156-171. doi: 10.4236/alamt.2014.43014.

Rhotrix Linear Transformation

Abdul Mohammed, Musa Balarabe, and Abdussamad Tanko Imam
Department of Mathematics, Ahmadu Bello University, Zaria, Nigeria

3

ABSTRACT

This paper considers rank of a rhotrix and characterizes its properties, as an extension of ideas to the rhotrix theory rhomboidal arrays, introduced in 2003 as a new paradigm of matrix theory of rectangular arrays. Furthermore, we present the necessary and sufficient condition under which a linear map can be represented over rhotrix.

INTRODUCTION

By a rhotrix A of dimension three, we mean a rhomboidal array defined as

$$A = \left\langle \begin{array}{ccc} & a & \\ b & c & d \\ & e & \end{array} \right\rangle,$$

Where, $a, b, c, d, e \in \Re$. The entry c in rhotrix A is called the heart of A and it is often denoted by h (A). The concept of rhotrix was introduced by [1] as an extension of matrix-tertions and matrix noitrets suggested by [2]. Since the introduction of rhotrix in [1], many researchers have shown interest on development of concepts for Rhotrix theory that are analogous to concepts in Matrix theory (see [3-9]). Sani [7] proposed an alternative method of rhotrix multiplication, by extending the concept of row-column multiplication of two dimensional matrices to three dimensional rhotrices, recorded as follows:

$$A \circ B = \left\langle \begin{matrix} & a & \\ b & h(A) & d \\ & e & \end{matrix} \right\rangle \circ \left\langle \begin{matrix} & f & \\ g & h(B) & i \\ & j & \end{matrix} \right\rangle$$

$$= \left\langle \begin{matrix} & af + dg & \\ bf + eg & h(A)h(B) & ai + dj \\ & bi + ej & \end{matrix} \right\rangle$$

Where, A and B belong to set of all three dimensional rhotrices, $R_3(\Re)$.

The definition of rhotrix was later generalized by [6] to include any finite dimension $n \in 2Z^+ + 1$. Thus; by a rhotrix A of dimension $n \in 2Z^+ + 1$ we mean a rhomboidal array of cardinality $\frac{1}{2}(n^2 + 1)$. Implying a rhotrix R of dimension n can be written as

$$R_n = \left\langle \begin{matrix} & & & a_{11} & & & \\ & & a_{21} & c_{11} & a_{12} & & \\ & \cdots & \cdots & \cdots & \cdots & \cdots & \\ a_{t1} & \cdots & \cdots & \cdots & \cdots & \cdots & a_{tt} \\ & \cdots & \cdots & \cdots & \cdots & \cdots & \\ & & a_{tt-1} & c_{t-1t-1} & a_{t-1t} & & \\ & & & a_{tt} & & & \end{matrix} \right\rangle$$

The element $a_{ij}(i.j = 1,2,\cdots,t)$ and $c_{kl}(k,l = 1,2,\cdots,t-1)$ are called the major and minor entries of R respectively. A generalization of row-column multiplication method for n-dimensional rhotrices was given by [8]. That is, given any n-dimensional rhotrices $R_n = \langle a_{ij}, c_{kl} \rangle$ and $Q_n = \langle b_{ij}, d_{kl} \rangle$, the multiplication of R_n and Q_n is as follows:

$$R_n \circ Q_n = \left\langle \sum_{i,j=1}^{t} \left(a_{ij} b_{ij}\right), \sum_{k,l=1}^{t-1} \left(c_{kl} d_{kl}\right) \right\rangle, t = \frac{(n+1)}{2}.$$

The method of converting a rhotrix to a special matrix called "coupled matrix" was suggested by [9]. This idea was used to solve systems of nxn and (n-1) × (n-1) matrix problems simultaneously. The concept of vectors and rhotrix vector spaces and their properties were introduced by [3] and [4] respectively. To the best of our knowledge, the concept of rank and linear transformation of rhotrix has not been studied. In this paper, we consider the rank of a rhotrix and characterize its properties. We also extend the idea to suggest the necessary and sufficient condition for representing rhotrix linear transformation.

PRELIMINARIES

The following definitions will help in our discussion of a useful result in this section and other subsequent ones.

Definition

Let $R_n = \langle a_{ij}, c_{kl} \rangle$ be an n-dimensional rhotrix. Then, a_{ij} is the (i,j)-entries called the major entries of R_n and c_{kl} is the (k,l)-entries called the minor entries of R_n.

Definition 2.2 [7]

A rhotrix $R_n = \langle a_{ij}, c_{kl} \rangle$ of n-dimension is a coupled of two matrices (a_{ij}) and (c_{kl}) consisting of its major and minor matrices respectively. Therefore, (a_{ij}) and (c_{kl}) are the major and minor matrices of R_n.

Definition

Let $R_n = \langle a_{ij}, c_{kl} \rangle$ be an n-dimensional rhotrix. Then, rows and columns of (a_{ij}) $((c_{kl}))$ will be called the major (minor) rows and columns of R_n respectively.

Definition

For any odd integer n, an n×n matrix (a_{ij}) is called a filled coupled matrix if a_{ij} =0 for all i,j whose sum i+j is odd. We shall refer to these entries as the null entries of the filled coupled matrix.

Theorem

There is one-one correspondence between the set of all n-dimensional rhotrices over F and the set of all n×n filled coupled matrices over F.

RANK OF A RHOTRIX

Let $R_n = \langle a_{ij}, c_{kl} \rangle$, the entries $a_{rr} (1 \le r \le t)$ and $c_{ss} (1 \le s \le t-1)$ in the main diagonal of the major and minor matrices of R respectively, formed the main diagonal of R. If all the entries to the left (right) of the main diagonal in R are zeros, Ris called a right (left) triangular rhotrix. The following lemma follows trivially.

Lemma

Let $R_n = \langle a_{ij}, c_{kl} \rangle$ is a left (right) triangular rhotrix if and only if (a_{ij}) and (C_{kl}) are lower (upper) triangular matrices.

Wang#title3_4: spProof

This follows when the rhotrix R_n is being rotated through 45° in anti-clockwise direction.

In the light of this lemma, any n-dimensional rhotrix R can be reduce to a right triangular rhotrix by reducing its major and minor matrix to echelon form using elementary row operations. Recall that, the rank of a matrix A denoted by rank (A) is the number of non-zero row(s) in its reduced row echelon form. If $R_n = \langle a_{ij}, c_{kl} \rangle$, we define rank of R denoted by rank (R) as:

$$\text{rank } (A) = \text{rank}(a_{ij}) + \text{rank}(c_{kl}) \tag{3}$$

It follows from equation (3) that many properties of rank of matrix can be extended to the rank of rhotrix. In particular, we have the following:

Theorem

Let $R_n = \langle a_{ij}, c_{kl} \rangle$ and $Q_n = \langle b_{ij}, d_{kl} \rangle$, be any two n-dimensional rhotrices, where $n \in 2Z^+ + 1$ Then

1. $rank(R) \leq n$;
2. $rank(R+S) \leq rank(R) + rank(S)$;
3. $rank(R) + rank(S) - n \leq rank(R \circ S)$;
4. $rank(R \circ S) \leq \min\{rank(R), rank(S)\}$.

Wang#title3_4: spProof

The first two statements follow directly from the definition. To prove the third statement, we apply the corresponding inequality for matrices, that is, $rank(AB) \geq rank(A) + rank(B) - n$ where A is m×n and B is n×p. Thus,

$$\begin{aligned}\operatorname{rank}(RS) &= \operatorname{rank}\left[(a_{ij})(b_{ij})\right] + \operatorname{rank}\left[(c_{kl})(d_{kl})\right] \\ &\geq \left[\operatorname{rank}(a_{ij}) + \operatorname{rank}(b_{ij}) - \left(\frac{n+1}{2}\right)\right] \\ &\quad + \left[\operatorname{rank}(c_{kl}) + \operatorname{rank}(d_{kl}) - \left(\frac{n+1}{2}\right) + 1\right] \\ &= \operatorname{rank}(R) + \operatorname{rank}(S) - n.\end{aligned}$$

For the last statement, consider

$$\begin{aligned}&\operatorname{rank}(RS) \\ &= \operatorname{rank}\left[(a_{ij})(b_{ij})\right] + \operatorname{rank}\left[(c_{kl})(d_{kl})\right] \\ &\leq \min\left\{(a_{ij}), \operatorname{rank}(b_{ij})\right\} + \min\left\{(c_{kl}), \operatorname{rank}(d_{kl})\right\} \\ &\leq \min\left\{(a_{ij}) + \operatorname{rank}(c_{kl}), (b_{ij}) + \operatorname{rank}(d_{kl})\right\} \\ &= \min\left\{\operatorname{rank}(R) + \operatorname{rank}(S)\right\}.\end{aligned}$$

Example

Let

$$A = \left\langle \begin{array}{ccccc} & & 1 & & \\ & 0 & 2 & -2 & \\ 1 & -1 & 3 & 1 & 2 \\ & -2 & 1 & 1 & \\ & & 2 & & \end{array} \right\rangle.$$

Then, the filled coupled matrix of A is given by

$$m(A)=\begin{pmatrix}1 & 0 & -2 & 0 & 2\\ 0 & 2 & 0 & 1 & 0\\ 0 & 0 & 3 & 0 & 1\\ 0 & -1 & 0 & 1 & 0\\ 1 & 0 & -2 & 0 & 2\end{pmatrix}.$$

Now reducing m (A) to reduce row echelon form (rref), we obtain

$$rref\left(m(A)\right)=\begin{pmatrix}1 & 0 & -2 & 0 & 2\\ 0 & 2 & 0 & 1 & 0\\ 0 & 0 & 3 & 0 & 1\\ 0 & 0 & 0 & 3 & 0\\ 0 & 0 & 0 & 0 & 0\end{pmatrix},$$

which is a coupled of (2×2) and (3×3) matrices, i.e.

$A(say)=\begin{pmatrix}2 & 0\\ 0 & 3\end{pmatrix}$ and $B(say)=\begin{pmatrix}1 & -2 & 2\\ 0 & 3 & 1\\ 0 & 0 & 0\end{pmatrix}$ respectively.

Notice that,

$\text{rank(A)} + \text{rank(B)} = 2+2 = 4 = \text{rank(rref(m(A)))}$

Hence, $= rank(A) = 4$.

RHOTRIX LINEAR TRANSFORMATION

One of the most important concepts in linear algebra is the concept of representation of linear mappings as matrices. If V and W are vector spaces of dimension n and m respectively, then any linear mapping T from V to W can be represented by a matrix. The matrix representation of T is called the matrix of T denoted by m (T). Recall that, if F is a field, then any vector space V of finite dimension n over F is isomorphic to F^n. Therefore, any n×n matrix over F can be considered as a linear operator on the vector space F^n in the fixed standard basis. Following this ideas, we study in this section, a rhotrix as a linear operator on the vector space F^n. Since the dimension of a rhotrix is always odd, it follow that, in representing a linear map T on a vector space V by a rhotrix, the dimension of V is necessarily odd. Therefore, throughout what follows, we shall consider only odd dimensional vector spaces. For any $n \in 2Z^+ + 1$ and F be an arbitrary field, we find the coupled F^t, F^{t-1} of F^t

$$F^t = \{(\alpha_1, \alpha_2, \cdots, \alpha_t) | \alpha_1, \cdots, \alpha_t \in F\}$$

And

$$F^{t-1} = \{(\beta_1, \beta_2, \cdots, \beta_t) | \beta_1, \beta_2, \cdots, \beta_{t-1} \in F^{t-1}\}$$

by

$$\begin{aligned}(F^t, F^{t-1}) = \{&(\alpha_1, \alpha_2, \cdots, \alpha_t, \beta_1, \beta_2, \cdots, \beta_{t-1}) : \\ &\alpha_1, \alpha_2, \cdots, \alpha_t, \beta_1, \beta_2, \cdots, \beta_{t-1} \in F^t\}.\end{aligned}$$

It is clear that (F^t, F^{t-1}) coincides with F^n and so, if $n \in 2Z^+ + 1$, any n-dimensional vector spaces V_1 and V_2 is of dimensions $\frac{n+1}{2}$ and $\frac{n+1}{2} - 1$ respectively. Less obviously, it can be seen that not every linear map T of F^n can be represented by a rhotrix in the standard basis. For instance, the map

$$T: F^3 \rightarrow F^3$$

defined by

$$T(x, y, z) = (x - y, x + z, y + z)$$

is a linear mapping on F^3 which cannot be represented by a rhotrix in the standard basis. The following theorem characterizes when a linear map T on F^n can be represented by a rhotrix.

Theorem

Let $n \in 2Z^+ + 1$ and F be a field. Then, a linear map $T: F^n \rightarrow F^n$ can be represented by a rhotrix with respect to the standard basis if and only if T is defined as

$$\begin{aligned}
&T(x_1, y_1, x_2, y_2, \cdots, y_{t-1}, x_t) \\
&= (\alpha_1(x_1, x_2, \cdots, x_t), \beta_1(y_1, y_2, \cdots, y_{t-1}), \\
&\quad \alpha_2(x_1, x_2, \cdots, x_t), \beta_2(y_1, y_2, \cdots, y_{t-1}), \cdots, \\
&\quad \beta_{t-1}(y_1, y_2, \cdots, y_{t-1}), \alpha_t(x_1, x_2, \cdots, x_t)),
\end{aligned}$$

where $t = \dfrac{n+1}{2}, \alpha_1, \cdots \alpha_t$ and $\beta_1, \cdots, \beta_{t-1}$ are any linear map on F^t and F^{t-1} respectively.

Proof

Suppose $T: F^n \rightarrow F^n$ is defined by

$$
\begin{aligned}
&T\left(x_1, y_1, x_2, y_2, \cdots, y_{t-1}, x_t\right) \\
&=\left(\alpha_1\left(x_1, x_2, \cdots, x_t\right), \beta_1\left(y_1, y_2, \cdots, y_{t-1}\right),\right. \\
&\quad \alpha_2\left(x_1, x_2, \cdots, x_t\right), \beta_2\left(y_1, y_2, \cdots, y_{t-1}\right), \cdots, \\
&\quad \left.\beta_{t-1}\left(y_1, y_2, \cdots, y_{t-1}\right), \alpha_t\left(x_1, x_2, \cdots, x_t\right)\right),
\end{aligned}
$$

where, $t=\dfrac{n+1}{2}, \alpha_1, \cdots, \alpha_t$ and $\beta_1, \cdots, \beta_{t-1}$ areanylinearmaponF^{t}andF^{t-1} respectively,andconsiderthestandardbasis $\{(1,0,\cdots,0),(0,1,0,\cdots,0),\cdots,(0,0,\cdots,1)\}$. Note that, for $1 \le i \le t$ and$1 \le j \le t-1$. Since α_i, β_j are linear maps, $\alpha_i(0,\cdots,0)=\beta_j(0,\cdots,0)=0$. Thus,

$$
\left.\begin{aligned}
&T(1,0,\cdots,0)=\left[\alpha_1(1,0,\cdots,0), 0, \cdots, \alpha_t(1,0,\cdots,0)\right] \\
&T(1,0,\cdots,0)=\left[0, \beta_1(1,0,\cdots,0), \cdots, \beta_{t-1}(1,0,\cdots,0)\right] \\
&\qquad\vdots \\
&T(0,\cdots,0,1)=\left[0, \beta_1(0,\cdots,0,1), \cdots, \beta_{t-1}(0,\cdots,0), 1\right] \\
&T(0,\cdots,0,1)=\left[\alpha_1(0,\cdots,0,1), 0, \cdots, \alpha_t(0,0,\cdots,0,1)\right]
\end{aligned}\right\} \tag{5}
$$

Let $\alpha_{ij}=\alpha_j\left(0,\cdots \underset{j^{th}-position}{1}, \cdots, 0\right)$ for $1 \le i, j \le t$ and $\beta_{kl}=\beta_l\left(0,\cdots \underset{j^{th}-position}{1}, \cdots, 0\right)$ for$1 \le k, l \le t-1$. Then from (5), we have the matrix of T is

$$
\begin{pmatrix}
\alpha_{11} & 0 & \alpha_{12} & \cdots & \alpha_{1t-1} & 0 & \alpha_{1t} \\
0 & \beta_{11} & 0 & \cdots & 0 & \beta_{1t-1} & 0 \\
\vdots & \vdots & \vdots & \ddots & \vdots & \vdots & \vdots \\
\vdots & \vdots & \vdots & \ddots & \vdots & \vdots & \vdots \\
\vdots & \vdots & \vdots & \ddots & \vdots & \vdots & \vdots \\
0 & \beta_{t-1t} & 0 & \cdots & 0 & \beta_{t-1t-1} & 0 \\
\alpha_{t1} & 0 & \alpha_{t2} & \cdots & \alpha_{tt-1} & 0 & \alpha_{tt}
\end{pmatrix}. \tag{6}
$$

This is a filled coupled matrix from which we obtain the rhotrix representation of T as $\langle \alpha_{ij}, \beta_{kl} \rangle$.

Conversely:

Suppose $T: F^n \to F^n$ has a rhotrix representation $\langle \alpha_{ij}, \beta_{kl} \rangle$ in the standard basis. Then, the corresponding matrix representation of T is the filled coupled given in (6) above. Thus, we obtain the system

$$\left.\begin{array}{l} T(1,0,\cdots,0) = (\alpha_{11}, 0, \alpha_{12}, \cdots, \alpha_{1t-1}, 0, \alpha_{1t}) \\ T(1,0,\cdots,0) = (0, \beta_{1t-1}, 0, \cdots, \beta_{1t-1}, 0) \\ \qquad\qquad\vdots \\ T(0,\cdots,0,1) = (0, \beta_{t-1t}, 0, \cdots, \beta_{t-1t-1}, 0) \\ T(0,\cdots,0,1) = (\alpha_{t1}, 0, \alpha_{t2}, \cdots, \alpha_{tt-1}, 0, \alpha_{tt}) \end{array}\right\} \tag{7}$$

From this system, it follows that for each $(x_1, y_1, x_2, y_2, \cdots, y_{t-1} x_1) \in F^n$ we have the linear transformation T defined by

$$\begin{aligned} & T(x_1, y_1, x_2, y_2, \cdots, y_{t-1}, x_t) \\ &= (\alpha_1(x_1, x_2, \cdots, x_t), \beta_1(y_1, y_2, \cdots, y_{t-1}), \\ &\quad \alpha_2(x_1, x_2, \cdots, x_t), \beta_2(y_1, y_2, \cdots, y_{t-1}), \cdots, \\ &\quad \beta_{t-1}(y_1, y_2, \cdots, y_{t-1}), \alpha_t(x_1, x_2, \cdots, x_t)), \end{aligned}$$

where, $t = \dfrac{n+1}{2}, \alpha_1, \cdots, \alpha_t$ and $\beta_1, \cdots, \beta_{t-1}$ are any linear map on F^t with

$\alpha_j\left(0, \cdots \underset{i^{th}-position}{1}, \cdots, 0\right) = \alpha_{ij}$ for $(1 \le i, j \le t)$ and $\beta_l\left(0, \cdots \underset{j^{th}-position}{1}, \cdots, 0\right) = \beta_{kl}$ for $(1 \le k, l \le t-1)$.

Example

Consider the linear mappings $T:\mathfrak{R}\rightarrow\mathfrak{R}$ define by $T(x,y,z)=(2x-z,4y,x-3z)$ to find the rhotrix of T relative to the standard basis. We proceed by finding the matrices of T. Thus,

$$T(1,0,0)=(2,0,1)$$
$$T(0,1,0)=(0,4,0)$$
$$T(0,0,1)=(-1,0,-3)$$

Therefore, by definition of matrix of T with respect to the standard basis, we have

$$m(T)=\begin{pmatrix}2 & 0 & 1\\ 0 & 4 & 0\\ -1 & 0 & -3\end{pmatrix},$$

which is a filled coupled matrix from which we obtain the rhotrix of T

in R_3, $r(T)=\left\langle\begin{matrix} & 2 & \\ -1 & 4 & 1\\ & -3 & \end{matrix}\right\rangle$.

Now starting with the rhotrix $r(T)=\left\langle\begin{matrix} & 2 & \\ -1 & 4 & 1\\ & -3 & \end{matrix}\right\rangle$ the filled coupled matrix

of r(T) is $\begin{pmatrix}2 & 0 & 1\\ 0 & 4 & 0\\ -1 & 0 & -3\end{pmatrix}$.

And so, defining $T:R_3\rightarrow R_3$

$$T(1,0,0)=2(1,0,0)+0(0,1,0)+1(0,0,1)$$
$$T(0,1,0)=0(1,0,0)+4(0,1,0)+0(0,0,1)$$
$$T(0,0,1)=-1(1,0,0)+0(0,1,0)-3(0,0,1)$$

Thus, if $(x,y,z)=x(1,0,0)+y(0,1,0)+z(0,0,1)$
Therefore,

$$\begin{aligned} T(x,y,z) &= xT(1,0,0)+yT(0,1,0)+zT(0,0,1) \\ &= x(2,0,1)+y(0,4,0)+z(-1,0,-3) \\ &= (2x-z,4y,x-3z) \end{aligned}$$

CONCLUSIONS

We have considered the rank of a rhotrix and characterize its properties as an extension of ideas to the rhotrix theory rhomboidal arrays. Furthermore, a necessary and sufficient condition under which a linear map can be represented over rhotrix had been presented.

ACKNOWLEDGEMENTS

The Authors wish to thank Ahmadu Bello University, Zaria, Nigeria for financial support towards publication of this article.

REFERENCES

1. A. O. Ajibade, "The Concept of Rhotrix in Mathematical Enrichment," International Journal of Mathematical Education in Science and Technology, Vol. 34, No. 2, 2003, pp. 175-179.doi:10.1080/0020739021000053828
2. K. T. Atanassov and A. G. Shannon, "Matrix-Tertions and Matrix-Noitrets: Exercise for Mathematical Enrichment," International Journal of Mathematical Education in Science and Technology, Vol. 29, No. 6, 1998, pp. 898–903.
3. A. Aminu, "Rhotrix Vector Spaces," International Journal of Mathematical Education in Science and Technology, Vol. 41, No. 4, 2010, pp. 531-573. doi:10.1080/00207390903398408
4. A. Aminu, "The Equation $R_n x = b$ over Rhotrices," International Journal of Mathematical Education in Science and Technology, Vol. 41, No. 1, 2010, pp. 98-105. doi:10.1080/00207390903189187

5. A. Mohammed, "Enrichment Exercises through Extension to Rhotrices," International Journal of Mathematical Education in Science and Technology, Vol. 38, No. 1, 2007, pp. 131-136. doi:10.1080/00207390600838490
6. A. Mohammed, "Theoretical Development and Applications of Rhotrices," Ph.D. Thesis, Ahmadu Bello University, Zaria, 2011.
7. B. Sani, "An Alternative Method for Multiplication of Rhotrices," International Journal of Mathematical Education in Science and Technology, Vol. 35, No. 5, 2004, pp. 777-781.doi:10.1080/00207390410001716577
8. B. Sani, "The Row-Column Multiplication of Higher Dimensional Rhotrices," International Journal of Mathematical Education in Science and Technology, Vol. 38, No. 5, 2007, pp. 657-662.
9. B. Sani, "Conversion of a Rhotrix to a 'Coupled Matrix'," International Journal of Mathematical Education in Science and Technology, Vol. 39, No. 2, 2008, pp. 244-249.doi:10.1080/00207390701500197

CITATION

1. A. Mohammed, M. Balarabe and A. Imam, "Rhotrix Linear Transformation," Advances in Linear Algebra & Matrix Theory, Vol. 2 No. 4, 2012, pp. 43-47. doi: 10.4236/alamt.2012.24007.

On Solving Centrosymmetric Linear Systems

Moawwad El-Mikkawy and Faiz Atlan
Mathematics Department, Faculty of Science,
Mansoura University, Mansoura, Egypt

4

ABSTRACT

The current paper is mainly devoted for solving centrosymmetric linear systems of equations. Formulae for the determinants of tridiagonal centrosymmetric matrices are obtained explicitly. Two efficient computational algorithms are established for solving general centrosymmetric linear systems. Based on these algorithms, a MAPLE procedure is written. Some illustrative examples are given.

INTRODUCTION

Throughout this paper, $|A|$ and A^T denote the determinant and the transpose of the matrix A respectively. Also $\lfloor x \rfloor$ denotes the greatest integer less than or equal to x Centrosymmetric matrices have practical applications in numerical analysis, information theory, statistics, physics, harmonic differential quadrature, differential equations, engineering, sinc methods, magic squares, linear system theory and pattern recognition. The interested readers may refer to [1-12].

Solving and analyzing linear systems of equations is a fundamental problem in science and engineering applications. The cost of solving any linear system using Gauss or Gauss-Gordan algorithms is $O(n^3)$ The motivation of the current paper is to develop efficient algorithms for solving any centrosymmetric linear system having n equations and n unknowns provided that the coefficient matrix of the system is nonsingular. The cost of each algorithm depends on the solvers of two associated linear systems having smaller sizes than n. More precisely, if $n = 2m$, then each of the two associated linear systems consists of m equations. If $n = 2m + 1$, then we have one system having m equations and the other has $(m + 1)$ equations. Consequently, if the two associated linear systems have special structures, then the cost of the centrosymmetric algorithm could be considerably reduced, in particular for large values of n.

The paper is organized as follows. In Section 2, some properties of the exchange and the rotate matrices are presented. Formulae for centrosymmetric tridiagonal determinants are obtained in Section 3. In Section 4, two computational algorithms for solving centrosymmetric linear systems are given. A MAPLE procedure is given in Section 5. Some illustrative examples are presented in Section 6.

Definition 1.1

The $n \times n$ matrix with 1's on the northeast-southeast diagonal and 0's elsewhere is called the exchange matrix of order n. The subscript on J_n is neglected whenever the size is obvious from the context.

$$J_n = \begin{bmatrix} 0 & L & L & 0 & 1 \\ M & & N & 1 & 0 \\ M & N & N & N & M \\ 0 & 1 & N & & M \\ 1 & 0 & L & L & 0 \end{bmatrix}, \qquad (1)$$

Definition 1.2

Let $A = (a_{ij})^n_{i,j=1}$ be a $n \times n$ matrix. The rotate of A, denoted A^R, is defined by

$$A^R = JAJ. \tag{2}$$

Definition 1.3 [9]

Let $A = (a_{ij})^n_{i,j=1}$ be a $n \times n$ matrix. Then

- A is said to be centrosymmetric if $A^R = A$.
- A is said to be per symmetric if $A^R = A^T$.
- A is said to be centrogonal if $A^R = A^{-1}$ provided $|A| \neq 0$.
- A is said to be skew-centrosymmetric if $A^R = -A$.
- A is said to be bisymmetric if $A^R = A = A^T$.

Note that the bisymmetric matrix is both symmetric and centrosymmetric. It is also both symmetric and presymmetric.

SOME PROPERTIES OF THE EXCHANGE AND THE ROTATE MATRICES

Let us begin this section by giving some helpful results concerning the exchange and rotate matrices. For more details see [1, 3, 8, 9, 12-21].

The exchange matrix J_n enjoys the following properties:

- $J = J^T = J^{-1}$.
- $J_n^k = \begin{cases} I_n, k \text{ even} \\ J_n, k \text{ odd}, \end{cases}$

- Where I_n is the identity matrix of order n.

- $|J_n| = (-1)^{n(n-1)/2} = \begin{cases} 1, \text{if } n \equiv 0 \text{ or } 1 \bmod(4) \\ -1, \text{if } n \equiv 2 \text{ or } 3 \bmod(4). \end{cases}$

- The matrix product JA is a version of the matrix A that has been reflected in line at 0 degree to the horizontal measured counter-clockwise.
- The matrix product AJ is a version of the matrix A that has been reflected in line at 90 degrees to the horizontal measured counter-clockwise.
- The matrix product $JAJ = A^R$ is a version of the matrix A that has been rotated counter-clockwise or clockwise by 180 degrees.

Note that the exchange matrix J is equal to the identity matrix I but with the columns in reverse order. More precisely, $J = (e_n, e_{n-1}, L, e_1)$ where $e_i = (\delta_{i1}, \delta_{i2}, L, \delta_{ni})$ $i = 1, 2, L, n,$. δ_{ij} is the Kronecker symbol which is equal to 1 if $i = j$ and zero if $i \neq j$.

It is also worth mentioned that the matrix J is sometimes called the counter-identity matrix or contra-identity matrix or per-identity matrix or the reflection matrix or the reversal matrix.

Exchange matrices are simple to construct in software platforms. For example, to construct J_5 in MAPLE, a single line of code can be used as follows:
n:= 5: J:= array(1..n,1..n,sparse): for i to n do J [i, n + 1 –i]:= 1 od: J Γ n:= op(J); or n:= 5: J:= matrix(n, n, 0): for i to n do J [i, n + 1 –i]:= 1 od: J Γ n:= op(J); The rotate of A, A^R of order n satisfies:

- $(A^R)^R = A$
- $(A^R)^T = (A^T)^R$
- $(A^{-1})^R = (A^R)^{-1}$, provided A^{-1} exists.
- $(AB)^R = A^R B^R$
- $(A \pm B)^R = (A^R \pm B^R)$

— $|A^R| = |A|$

— If row_i of A is $(a_{i1}, a_{i2}, L, a_{i,n-1}, a_{in})$, then row_{n+1-i}

of A^R is $(a_{in}, a_{i,n-1}, L, a_{i2}, a_{i1})$ In other words, the centrosymmetric matrix A is the same when read backwards as when read forwards.

CENTROSYMMETRIC DETERMINANTS

Centrosymmetric determinants take the form:

$$\begin{vmatrix} a_{1,1} & a_{1,2} & L & a_{1,n} \\ a_{2,1} & a_{2,2} & L & a_{2,n} \\ M & M & M & M \\ a_{2,n} & L & a_{2,2} & a_{2,1} \\ a_{1,n} & L & a_{1,2} & a_{1,1} \end{vmatrix}, \tag{3}$$

In which $\text{row}_i = (a_{i1}, a_{i2}, L, a_{i,n-1}, a_{in})$ and $\text{row}_{n+1-i} = (a_{in}, a_{i,n-1}, L, a_{i2}, a_{i1})$

for each $i = 1, 2, L, \left\lfloor \frac{n}{2} \right\rfloor$.

In particular, centrosymmetric tridiagonal determinants are of special importance. For convenience of the reader, we present some definitions, notations and properties associated with tridiagonal matrices.

A tridiagonal matrix $T = (a_{ij})_{i,j=1}^{n}$ takes the form:

$$T = \begin{bmatrix} d_1 & a_1 & 0 & L & 0 \\ b_2 & d_2 & a_2 & O & M \\ 0 & O & O & O & 0 \\ M & O & b_{n-1} & d_{n-1} & a_{n-1} \\ 0 & L & 0 & b_n & d_n \end{bmatrix}. \tag{4}$$

These types of matrices frequently appear in many areas of science and engineering. For example in parallel computing, telecommunication system analysis and in solving differential equations using finite differences, see [22, 23]. A general $n \times n$ tridiagonal matrix of the form (4) can be stored in 3n memory locations, rather than n^2 memory locations for a full matrix, by using three vectors a= (a_1, a_2, L, a_n), b= (b_1, b_2, L, b_n), and d= (d_1, d_2, L, d_n) with b_1=a_n=0 This is always a good habit in computation in order to save memory space. To study tridiagonal matrices it is very convenient to introduce a vector c in the following way [24]:

$$c = \left(c_1, c_2, \cdots, c_n\right), \tag{5}$$

Where

$$c_1 = d_1, \quad c_k = d_k - \frac{b_k a_{k-1}}{c_{k-1}}, \quad k = 2, 3, \cdots, n. \tag{6}$$

It is helpful to restate some important results concerning tridiagonal matrices of the form (4). For more details the interested reader may refer to [24-28].

Let p, q, r and s are positive integers such that $1 \le p \le q \le n$ and $1 \le r \le s \le n$.

Define the sub matrix of A, denoted A[p..q,r..s] of order $(q-p+1) \times (s-r+1)$ by

$$A\left[p..q, r..s\right] = \begin{bmatrix} a_{p,r} & a_{p,r+1} & \mathrm{L} & a_{p,s} \\ a_{p+1,r} & a_{p+1,r+1} & \mathrm{L} & a_{p+1,s} \\ \mathrm{M} & \mathrm{M} & \mathrm{O} & \mathrm{M} \\ a_{q,r} & a_{q,r+1} & \mathrm{L} & a_{q,s} \end{bmatrix}. \tag{7}$$

In particular, let $F_k = A[1..k, 1..k]$, $1 \le k \le n$ and $f_k = |F_k|$, the leading principal minor of A.

Theorem 3.1 [24]

Consider

$$f_1 = \left|T[1..1,1..1]\right| = \left|d_1\right| = d_1,$$
$$f_k = \left|T[1..k,1..k]\right| = \left|F_k\right|,\ k = 2,3,\cdots,n. \tag{8}$$

Then the determinants in (8) satisfy a three-term recurrence

$$f_k = d_k\, f_{k-1} - b_k a_{k-1}\, f_{k-2},\ k = 1,2,\cdots,n, \tag{9}$$

Where the initial values for f_k are $f_0 = 1$ and $f_{-1} = 0$.

Lemma 3.2 [29]

If the LU factorization of the matrix T in (4) is possible, then we have

$$|T| = \prod_{k=1}^{n} c_k. \tag{10}$$

Where c_1, c_2, L, c_n are given by (6). Meanwhile, the Doolittle LU factorization [30] of T is given by:

$$T = LU, \text{ where}$$
$$L = \begin{bmatrix} 1 & 0 & L & L & 0 \\ \frac{b_2}{c_1} & 1 & O & & M \\ 0 & O & O & O & 0 \\ M & O & \frac{b_{n-1}}{c_{n-2}} & 1 & 0 \\ 0 & L & 0 & \frac{b_n}{c_{n-1}} & 1 \end{bmatrix},$$
$$\text{and } U = \begin{bmatrix} c_1 & a_1 & 0 & L & 0 \\ 0 & c_2 & a_2 & O & M \\ M & O & O & O & 0 \\ M & & O & c_{n-1} & a_{n-1} \\ 0 & L & L & 0 & c_n \end{bmatrix}. \tag{11}$$

Lemma 3.3 [26]

If $c_r \neq 0, r = 1, 2, \cdots, n$ then

$$f_i = \prod_{k=1}^{i} c_k, \; i = 1, 2, \cdots, n. \tag{12}$$

Lemma 3.4 [26]

If $c_i = 0,\; 1 \leq i \leq n-1$, and either $a_i = 0$ or $b_{i+1} = 0$, then T is a singular matrix.

Lemma 3.5 [24]

If $c_i \neq 0$ for each i= 1, 2, $\cdots$, n-1, then the three-term recurrence (9) reduces to the two term recurrence

$$f_i = c_i \, f_{i-1}, \quad i = 1, 2, \cdots, n. \tag{13}$$

Algorithm 3.1 (DETGTRI [31]).

The determinant of the matrix in (4) can be computed using the following symbolic algorithm.

INPUT: Order of the matrix n and the components, a_i, d_i, b_i, i=1, 2, $\cdots$, n, (a_n=b_1=0).

OUTPUT: The determinant of the matrix T in (4).

Step 1: Use (6) to compute the simplest forms of the n components of the vector c. If c_i=0 for any $i \leq n$, set c_i=t (t is just a symbolic name) and continue to compute $c_{i+1}, c_{i+2}, \cdots, c_n$ in terms of t by using (6).

Step 2: The simplest rational form of the product $P(t) = \prod_{r=1}^{n} c_r$ (This product is a polynomial in t) evaluated at t=0 is equal to the determinant of the matrix T in (4), i.e. $|T| = P(0)$.

The cost of the DETGTRI algorithm is O (n). The algorithm is easy to implement in all Computer Algebra Systems (CAS) such as MACSYMA, MATHEMATICA and MAPLE...

Lemma 3.6

Consider the tridiagonal matrix T_n given by:

$$T_n = \begin{bmatrix} d_1 & a_1 & 0 & \mathrm{L} & 0 \\ b_2 & d_2 & a_2 & \mathrm{O} & \mathrm{M} \\ 0 & \mathrm{O} & \mathrm{O} & \mathrm{O} & 0 \\ \mathrm{M} & \mathrm{O} & b_{n-1} & d_{n-1} & a_{n-1} \\ 0 & \mathrm{L} & 0 & b_n & d_n \end{bmatrix}. \tag{14}$$

Let S_n, U_n and V_n be $n \times n$ matrices defined respectively as follows:

$$S_n = T_n\big|_{d_n \to d_n \pm k}, \quad U_n = T_n\big|_{a_{n-1} \to k a_{n-1}}$$
$$\text{and} \quad V_n = T_n\big|_{b_n \to k b_n}, \tag{15}$$

Where k is a scalar quantity. Then by applying the DETGTRI algorithm, we see that: having used (12) and (13).

$$|S_n| = |T_n| \pm k|T_{n-1}|, \tag{16}$$

$$|U_n| = |V_n| = |T_n| - (k-1)\, b_n a_{n-1} |T_{n-2}|, \tag{17}$$

Let $A=(a_{ij})_{i,j=1}^{n}$ is a $n\times n$ centrosymmetric matrix, then the three following facts are useful when we deal with such matrices:

Fact (1): If $row_i=(a_{i1},a_{i2},L,a_{i,n-1},a_{in})$ then $row_{n+1-i}=(a_{in},a_{i,n-1},L,a_{i2},a_{i1})$. In other words, the centrosymmetric matrix A is the same when read backwards as when read forwards.

Fact (2): If n=2m is positive even number, then $Q_1=1/\sqrt{2}\begin{bmatrix} I_m & I_m \\ J_m & -J_m \end{bmatrix}$ is orthogonal.

Fact (3): If n=2m+1 is positive odd number then $Q_2=1/\sqrt{2}\begin{bmatrix} I_m & 0 & I_m \\ 0 & \sqrt{2} & 0 \\ J_m & 0 & -J_m \end{bmatrix}$ is orthogonal. Armed with the above facts, we may formulate the following result whose proof will be omitted.

Theorem 3.7

Let $R=(r_{ij})_{i,j=1}^{n}$ be an $n\times n$ of even order say n=2m then R can be written in the form:

$$R=\begin{bmatrix} A & JBJ \\ B & JAJ \end{bmatrix}=Q_1\begin{bmatrix} A+JB & \\ & A-JB \end{bmatrix}Q_1^T, \tag{18}$$

Where $A,B\in i^{m\times m}$ The determinant of the matrix R in (18) is given by:

$$|R|=|A+JB||A-JB|. \tag{19}$$

If n is odd, i.e., n=2m+1, then we have:

$$R=\begin{bmatrix} A & v & JBJ \\ u^{\mathrm{T}} & q & u^{\mathrm{T}}J \\ B & Jv & JAJ \end{bmatrix}$$

$$=Q_2\begin{bmatrix} A+JB & \sqrt{2}v & 0 \\ \sqrt{2}u^{\mathrm{T}} & q & 0 \\ 0 & 0 & A-JB \end{bmatrix}Q_2^{\mathrm{T}}. \tag{20}$$

Where $A,B \in i^{m\times m}$, $u,v \in i^{m\times 1}$ and $q \in i$. The determinant of the matrix R in (20) is given by

$$|R|=\begin{vmatrix} A+JB & \sqrt{2}v \\ \sqrt{2}u^{\mathrm{T}} & q \end{vmatrix}|A-JB|. \tag{21}$$

Concerning the inverse of centrosymmetric matrices the reader may refer to [8]. As an interesting special case of the Theorem 3.7, we give the following result.

Corollary 3.8

In Theorem 3.7, if $R = T_n$ is centrosymmetric tridiagonal matrix, then we have

$$|T_n|=\begin{cases} |T_m|^2-k^2|T_{m-1}|^2, & \text{if } n=2m, \\ |T_m|\left(|T_{m+1}|-b_{m+1}a_m|T_{m-1}|\right), & \text{if } n=2m+1. \end{cases}$$

Where k is the common value of the elements in positions (m, m+1) and (m+1, m) of the matrix $R = T_n$ when n = 2m.

Proof: To compute the centrosymmetric tridiagonal determinant of order n. Two cases will be considered:

Case (1): n=2m In this case, the centrosymmetric tridiagonal matrix takes the form:

$$T_n = T_{2m} = \left[\begin{array}{ccccc|ccccc}
d_1 & a_1 & 0 & \cdots & 0 & 0 & \cdots & & \cdots & 0 \\
b_2 & d_2 & a_2 & \ddots & \vdots & \vdots & & & & \vdots \\
0 & \ddots & \ddots & \ddots & 0 & & & & & \\
\vdots & \ddots & b_{m-1} & d_{m-1} & a_{m-1} & 0 & & & & \vdots \\
0 & \cdots & 0 & b_m & d_m & k & 0 & & \cdots & 0 \\
\hline
0 & \cdots & & 0 & k & d_m & b_m & 0 & \cdots & 0 \\
\vdots & & & & 0 & a_{m-1} & d_{m-1} & b_{m-1} & \ddots & \vdots \\
 & & & & & 0 & \ddots & \ddots & \ddots & 0 \\
\vdots & & & & \vdots & \vdots & \ddots & a_2 & d_2 & b_2 \\
0 & \cdots & & \cdots & 0 & 0 & \cdots & 0 & a_1 & d_1
\end{array}\right]. \qquad (22)$$

Applying Theorem 3.7, we have $|T_n| = |T_{2m}| = |L||M|$, where

$$L = A + JB = \begin{bmatrix} d_1 & a_1 & 0 & \cdots & 0 \\ b_2 & d_2 & a_2 & \ddots & \vdots \\ 0 & \ddots & \ddots & \ddots & 0 \\ \vdots & \ddots & b_{m-1} & d_{m-1} & a_{m-1} \\ 0 & \cdots & 0 & b_m & d_m + k \end{bmatrix}$$

And

$$M = A - JB = \begin{bmatrix} d_1 & a_1 & 0 & \cdots & 0 \\ b_2 & d_2 & a_2 & \ddots & \vdots \\ 0 & \ddots & \ddots & \ddots & 0 \\ \vdots & \ddots & b_{m-1} & d_{m-1} & a_{m-1} \\ 0 & \cdots & 0 & b_m & d_m - k \end{bmatrix}.$$

By using Lemma 3.6, we obtain $|L| = |T_m| + k|T_{m-1}|$ and $|M| = |T_m| - k|T_{m-1}|$. Therefore we get

$$
\begin{aligned}
|T_n| &= |T_{2m}| = |L||M| \\
&= |T_m|^2 - k^2 |T_{m-1}|^2 .
\end{aligned}
\tag{23}
$$

Case (2): n=2m+1 In this case, the centrosymmetric tridiagonal matrix takes the form:

$$
T_n = T_{2m+1} = \left[\begin{array}{ccccc|c|ccccc}
d_1 & a_1 & 0 & L & 0 & 0 & 0 & L & & L & 0 \\
b_2 & d_2 & a_2 & O & M & M & M & & & & M \\
0 & O & O & O & 0 & M & & & & & \\
M & O & b_{m-1} & d_{m-1} & a_{m-1} & 0 & M & & & & M \\
0 & L & 0 & b_m & d_m & a_m & 0 & L & & L & 0 \\
\hline
0 & L & L & 0 & b_{m+1} & d_{m+1} & b_{m+1} & 0 & L & & 0 \\
\hline
0 & L & & L & 0 & a_m & d_m & b_m & 0 & L & 0 \\
M & & & & M & 0 & a_{m-1} & d_{m-1} & b_{m-1} & O & M \\
 & & & & & M & 0 & O & O & O & 0 \\
M & & & & M & M & M & O & a_2 & d_2 & b_2 \\
0 & L & & L & 0 & 0 & 0 & L & 0 & a_1 & d_1
\end{array}\right].
\tag{24}
$$

Applying Theorem 3.7, we obtain $|T_n| = |T_{2m+1}| = |L||M|$ here

$$
L = A - JB = A = T_m = \begin{bmatrix}
d_1 & a_1 & 0 & L & 0 \\
b_2 & d_2 & a_2 & O & M \\
0 & O & O & O & 0 \\
M & O & b_{m-1} & d_{m-1} & a_{m-1} \\
0 & L & 0 & b_m & d_m
\end{bmatrix}
$$

And

$$
M = \begin{bmatrix} A + JB & \sqrt{2}\boldsymbol{v} \\ \sqrt{2}\boldsymbol{u}^{\mathrm{T}} & q \end{bmatrix} = \begin{bmatrix}
d_1 & a_1 & 0 & L & 0 \\
b_2 & d_2 & a_2 & O & M \\
0 & O & O & O & 0 \\
M & O & b_m & d_m & \sqrt{2}a_m \\
0 & L & 0 & \sqrt{2}b_{m+1} & d_{m+1}
\end{bmatrix}.
$$

By using Lemma 3.6, we obtain $|\mathrm{L}| = |\mathrm{T_m}|$ and

$$|M| = \begin{vmatrix} A+JB & \sqrt{2}\boldsymbol{v} \\ \sqrt{2}\boldsymbol{u}^{\mathrm{T}} & q \end{vmatrix} = \begin{vmatrix} d_1 & a_1 & 0 & \mathrm{L} & 0 \\ b_2 & d_2 & a_2 & \mathrm{O} & \mathrm{M} \\ 0 & \mathrm{O} & \mathrm{O} & \mathrm{O} & 0 \\ \mathrm{M} & \mathrm{O} & b_m & d_m & \sqrt{2}a_m \\ 0 & \mathrm{L} & 0 & \sqrt{2}b_{m+1} & d_{m+1} \end{vmatrix}.$$

By using the DETGTRI algorithm [31] together with (12) and (13), then we have

$$|M| = \begin{vmatrix} d_1 & a_1 & 0 & \mathrm{L} & 0 \\ b_2 & d_2 & a_2 & \mathrm{O} & \mathrm{M} \\ 0 & \mathrm{O} & \mathrm{O} & \mathrm{O} & 0 \\ \mathrm{M} & \mathrm{O} & b_m & d_m & 2a_m \\ 0 & \mathrm{L} & 0 & b_{m+1} & d_{m+1} \end{vmatrix} = \begin{vmatrix} d_1 & a_1 & 0 & \mathrm{L} & 0 \\ b_2 & d_2 & a_2 & \mathrm{O} & \mathrm{M} \\ 0 & \mathrm{O} & \mathrm{O} & \mathrm{O} & 0 \\ \mathrm{M} & \mathrm{O} & b_m & d_m & a_m \\ 0 & \mathrm{L} & 0 & 2b_{m+1} & d_{m+1} \end{vmatrix}$$

$$= (c_1 c_2 \mathrm{L}\ c_m)\alpha_{m+1} = |T_m|\alpha_{m+1} = |T_m|\left(d_{m+1} - 2\frac{b_{m+1}a_m}{c_m}\right) = |T_m|\left(c_{m+1} - \frac{b_{m+1}a_m}{c_m}\right)$$

$$= c_{m+1}|T_m| - \frac{|T_m|}{c_m}b_{m+1}a_m = |T_{m+1}| - \frac{|T_m|}{c_m}b_{m+1}a_m = |T_{m+1}| - b_{m+1}a_m|T_{m-1}|.$$

Therefore

$$|T_n| = |T_{2m+1}|$$
$$= |T_m|\left(|T_{m+1}| - b_{m+1}\,a_m|T_{m-1}|\right). \tag{25}$$

From (23) and (25), we see that in order to compute the determinant of a centrosymmetric tridiagonal matrix of order n, then all we need is to compute c_1, c_2, L, c_m if n=2m and c_1, c_2, L, c_{m+1} if n=2m+1.

ALGORITHMS FOR SOLVING CENTROSYMMETRIC LINEAR SYSTEMS

Solving linear systems practically dominates scientific computing. In the present section, we focus on solving linear systems of centrosymmetric type. Two cases will be considered:

Case (i): n=2m. For this case we are going to construct an algorithm for solving centrosymmetric linear systems of the form:

$$\begin{bmatrix} a_{11} & a_{12} & \mathrm{L} & a_{1,m} & a_{1,m+1} & \mathrm{L} & a_{1,2m-1} & a_{1,2m} \\ a_{21} & a_{22} & \mathrm{L} & a_{2,m} & a_{2,m+1} & \mathrm{L} & a_{2,2m-1} & a_{2,2m} \\ \mathrm{M} & \mathrm{M} & \mathrm{O} & \mathrm{M} & \mathrm{M} & \mathrm{O} & \mathrm{M} & \mathrm{M} \\ a_{m1} & a_{m2} & \mathrm{L} & a_{m,m} & a_{m,m+1} & \mathrm{L} & a_{m,2m-1} & a_{m,2m} \\ a_{m,2m} & a_{m,2m-1} & \mathrm{L} & a_{m,m+1} & a_{m,m} & \mathrm{L} & a_{m,2} & a_{m,1} \\ \mathrm{M} & \mathrm{M} & \mathrm{O} & \mathrm{M} & \mathrm{M} & \mathrm{O} & \mathrm{M} & \mathrm{M} \\ a_{2,2m} & a_{2,2m-1} & \mathrm{L} & a_{2,m+1} & a_{2,m} & \mathrm{L} & a_{22} & a_{21} \\ a_{1,2m} & a_{1,2m-1} & \mathrm{L} & a_{1,m+1} & a_{1,m} & \mathrm{L} & a_{12} & a_{11} \end{bmatrix} \begin{bmatrix} x_1 \\ x_2 \\ \mathrm{M} \\ x_m \\ \mathrm{M} \\ x_{2m-1} \\ x_{2m} \end{bmatrix} = \begin{bmatrix} b_1 \\ b_2 \\ \mathrm{M} \\ b_m \\ \mathrm{M} \\ b_{2m-1} \\ b_{2m} \end{bmatrix}. \tag{26}$$

Block multiplication is particularly useful when there are patterns in the matrices to be multiplied. Therefore it is convenient to rewrite (26) in the partitioned form

$$\left[\begin{array}{c|c} A & JBJ \\ \hline B & JAJ \end{array}\right] \left[\begin{array}{c} X_1 \\ \hline X_2 \end{array}\right] = \left[\begin{array}{c} B_1 \\ \hline B_2 \end{array}\right], \tag{27}$$

Where

$$A = \begin{bmatrix} a_{11} & a_{12} & \mathrm{L} & a_{1,m} \\ a_{21} & a_{22} & \mathrm{L} & a_{2,m} \\ \mathrm{M} & \mathrm{M} & \mathrm{O} & \mathrm{M} \\ a_{m1} & a_{m2} & \mathrm{L} & a_{m,m} \end{bmatrix} \quad B = \begin{bmatrix} a_{m,2m} & a_{m,2m-1} & \mathrm{L} & a_{m,m+1} \\ \mathrm{M} & \mathrm{M} & \mathrm{O} & \mathrm{M} \\ a_{2,2m} & a_{2,2m-1} & \mathrm{L} & a_{2,m+1} \\ a_{1,2m} & a_{1,2m-1} & \mathrm{L} & a_{1,m+1} \end{bmatrix},$$

$$X_1 = [x_1, x_2, \ldots, x_m]^{\mathrm{T}},\ X_2 = [x_{m+1}, x_{m+2}, \ldots, x_{2m}]^{\mathrm{T}},$$

$$B_1 = [b_1, b_2, \ldots, b_m]^{\mathrm{T}} \text{ and } B_2 = [b_{m+1}, b_{m+2}, \ldots, b_{2m}]^{\mathrm{T}}.$$

The system in (26) can also be written in matrix form as follows:

$$R\boldsymbol{x} = \boldsymbol{b}, \tag{28}$$

Where $R = (a_{ij})_{i,j=1}^{2m}$ is the coefficient matrix of the system (26), $x = [x_1, x_2, \ldots, x_{2m}]^{\mathrm{T}}$ and $b = [b_1, b_2, \ldots, b_{2m}]^{\mathrm{T}}$ is the constant vector.

Algorithm 4.1

An algorithm for solving centrosymmetric linear system of even order. To solve the linear system of the form (26), we may proceed as follows:

INPUT: The entries of the coefficient matrix R and the constant vector b in (28).

OUTPUT: Solution vector $x = [x_1, x_2, \ldots, x_{2m}]^{\mathrm{T}}$.

Step 1: Construct the $m \times m$ matrices P, Q and the m-vectors $\hat{b}$ and $\tilde{b}$ as follows:

$$P = A + JB = \left[a_{ij} + a_{i,2m+1-j}\right]_{i,j=1}^{m},$$

$$Q = A - JB = \left[a_{ij} - a_{i,2m+1-j}\right]_{i,j=1}^{m},$$

$$\hat{\boldsymbol{b}} = [b_1 + b_{2m}, b_2 + b_{2m-1}, \ldots, b_m + b_{m+1}]^{\mathrm{T}}$$

and

$$\tilde{\boldsymbol{b}} = [b_1 - b_{2m}, b_2 - b_{2m-1}, \cdots, b_m - b_{m+1}]^{\mathrm{T}}.$$

Step 2: Compute $|R| = |P||Q|$ If $|R| = 0$ then Exiterror ('No solutions') end if.

Step 3: Solve the two linear systems:

$Py = \hat{b}$, and $Qz = \tilde{b}$ for $y = [y_1, y_2, \ldots, y_m]^T$ and $z = [z_1, z_2, \ldots, z_m]^T$ respectively.

Step 4: The solution vector $x = [x_1, x_2, \ldots, x_{2m}]^T$ is given by

$$x_i = \begin{cases} \frac{1}{2}(y_i + z_i) & \text{if } i = 1, 2, \ldots, m, \\ \frac{1}{2}(y_{2m+1-i} - z_{2m+1-i}) & \text{if } i = m+1, m+2, \ldots, 2m. \end{cases}$$

The Algorithm 4.1, will be referred to as CENTROSYMM-I algorithm. Concerning the computational cost of the CENTROSYMM-I algorithm:

1. The time complexity of Step 1 is
   $$m^2 + m^2 + m + m$$
   $$= 2m^2 + 2m = 2 \times \left(\frac{n}{2}\right)^2 + 2 \times \frac{n}{2}$$
   $$= \frac{1}{2}n^2 + n = \frac{1}{2}n^2 + O(n)$$
   ($\frac{1}{2}n^2 + n$ additions/subtractions and no multiplications/divisions).
2. The time complexity of Step 3 depends on the solvers of the two linear systems. For example, tridiagonal linear systems can be solved in linear time (see [25]).
3. The time complexity of Step 4 is O (n) (n additions/subtractions and n multiplications).

Step 3 is the step that leads to the reduction of the time complexity, because instead of solving a linear system of n equations, we end up with two linear systems half the size of the original one. If the original system is solved with Gaussian elimination (GE) method, then the time complexity will be $\frac{2}{3}n^3 + O(n^2)$. But, if GE method is used to solve the two systems in the third step, then the time complexity of our algorithm will be $\frac{4}{3}m^3 + O(m^2) = \frac{6}{3}n^3 + O(n^2)$ which is a significant reduction. If a method more efficient than the GE method is used, then the time complexity of our algorithm will be less (see also [16]).

Case (ii): n=2m+1 In this case the linear system to be considered has the form:

$$\begin{bmatrix} a_{11} & a_{12} & L & a_{1,m} & a_{1,m+1} & a_{1,m+2} & L & a_{1,2m} & a_{1,2m+1} \\ a_{21} & a_{22} & L & a_{2,m} & a_{2,m+1} & a_{2,m+2} & L & a_{2,2m} & a_{2,2m+1} \\ M & M & O & M & M & M & O & M & M \\ a_{m1} & a_{m2} & L & a_{m,m} & a_{m,m+1} & a_{m,m+2} & L & a_{m,2m} & a_{m,2m+1} \\ a_{m+1,1} & a_{m+1,2} & L & a_{m+1,m} & a_{m+1,m+1} & a_{m+1,m} & L & a_{m+1,2} & a_{m+1,1} \\ a_{m,2m+1} & a_{m,2m} & L & a_{m,m+2} & a_{m,m+1} & a_{m,m} & L & a_{m,2} & a_{m,1} \\ M & M & O & M & M & M & O & M & M \\ a_{2,2m+1} & a_{2,2m} & L & a_{2,m+2} & a_{2,m+1} & a_{2,m} & L & a_{22} & a_{21} \\ a_{1,2m+1} & a_{1,2m} & L & a_{1,m+2} & a_{1,m+1} & a_{1,m} & L & a_{12} & a_{11} \end{bmatrix} \begin{bmatrix} x_1 \\ x_2 \\ M \\ x_{m+1} \\ M \\ x_{2m} \\ x_{2m+1} \end{bmatrix} = \begin{bmatrix} b_1 \\ b_2 \\ M \\ b_{m+1} \\ M \\ b_{2m} \\ b_{2m+1} \end{bmatrix}, \qquad (29)$$

Or equivalently,

$$\left[\begin{array}{c|c|c} A & v & JBJ \\ \hline u^T & q & u^T J \\ \hline B & Jv & JAJ \end{array}\right] \left[\begin{array}{c} X_1 \\ \hline x_{m+1} \\ \hline X_2 \end{array}\right] = \left[\begin{array}{c} B_1 \\ \hline b_{m+1} \\ \hline B_2 \end{array}\right], \qquad (30)$$

Where

$$A = \begin{bmatrix} a_{11} & a_{12} & L & a_{1,m} \\ a_{21} & a_{22} & L & a_{2,m} \\ M & M & O & M \\ a_{m1} & a_{m2} & L & a_{m,m} \end{bmatrix} \quad B = \begin{bmatrix} a_{m,2m+1} & a_{m,2m} & L & a_{m,m+2} \\ M & M & O & M \\ a_{2,2m+1} & a_{2,2m} & L & a_{2,m+2} \\ a_{1,2m+1} & a_{1,2m} & L & a_{1,m+2} \end{bmatrix},$$

$$q = \left[a_{m+1,m+1} \right],\ \boldsymbol{u} = \left[a_{m+1,1}, a_{m+1,2}, \cdots, a_{m+1,m} \right]^{\mathrm{T}},\ \boldsymbol{v} = \left[a_{1,m+1}, a_{2,m+1}, \cdots, a_{m,m+1} \right]^{\mathrm{T}},$$

$$X_2 = \left[x_{m+2}, x_{m+3}, \cdots, x_{2m+1} \right]^{\mathrm{T}},\ X_1 = \left[x_1, x_2, \cdots, x_m \right]^{\mathrm{T}},$$

$$B_1 = \left[b_1, b_2, \cdots, b_m \right]^{\mathrm{T}} \text{ and } B_2 = \left[b_{m+2}, b_{m+3}, \cdots, b_{2m+1} \right]^{\mathrm{T}}.$$

The linear system (30) can also be written as:

$$R\boldsymbol{x} = \boldsymbol{b}, \tag{31}$$

Where $R = (a_{ij})_{i,j=1}^{2m+1}$ is the coefficient matrix of the system (29), $x = [x_1, x_2, \cdots, x_{2m+1}]^{\mathrm{T}}$ and $b = [b_1, b_2, \cdots, b_{2m+1}]^{\mathrm{T}}$ is the constant vector.

Algorithm 4.2

An algorithm for solving centrosymmetric linear system of odd order. To solve the linear system of the form (29), we may proceed as follows:

INPUT: The entries of the coefficient matrix R and the constant vector b in (31).

OUTPUT: Solution vector $x = [x_1, x_2, \cdots, x_{2m+1}]^{\mathrm{T}}$.

Step 1: Construct the matrices P, Q of orders m+1 and m respectively and the vectors $\hat{b}$ and $\tilde{b}$ of dimensions m+1 and m respectively as follows:

$$P = \begin{bmatrix} A + JB & 2\boldsymbol{v} \\ \boldsymbol{u}^{\mathrm{T}} & q \end{bmatrix} = \begin{bmatrix} a_{1,1} + a_{1,2m+1} & a_{1,2} + a_{1,2m} & \cdots & a_{1,m} + a_{1,m+2} & 2a_{1,m+1} \\ a_{2,1} + a_{2,2m+1} & a_{2,2} + a_{2,2m} & \cdots & a_{2,m} + a_{2,m+2} & 2a_{2,m+1} \\ \vdots & \vdots & \ddots & \vdots & \vdots \\ a_{m,1} + a_{m,2m+1} & a_{m,2} + a_{m,2m} & \cdots & a_{m,m} + a_{m,m+2} & 2a_{m,m+1} \\ a_{m+1,1} & a_{m+1,2} & \cdots & a_{m+1,m} & a_{m+1,m+1} \end{bmatrix},$$

$$Q = A - JB = \left[a_{ij} - a_{i,2m+2-j} \right]_{i,j=1}^{m},$$

$$\hat{\boldsymbol{b}} = \left[b_1 + b_{2m+1}, b_2 + b_{2m}, \cdots, b_m + b_{m+2}, b_{m+1} \right]^{\mathrm{T}}$$

and

$$\tilde{\boldsymbol{b}} = \left[b_1 - b_{2m+1}, b_2 - b_{2m}, \cdots, b_m - b_{m+2} \right]^{\mathrm{T}}.$$

Step 2: Compute $|R| = |P||Q|$, If $|R| = 0$ then Exiterror ("No solutions") end if.

Step 3: Solve the two linear systems $\mathrm{P}y = \hat{\mathrm{b}}$ and $\mathrm{Q}z = \tilde{\mathrm{b}}$ for $y = [y_1, y_2, \cdots, y_m, y_{m+1}]^{\mathrm{T}}$ and $z = [z_1, z_2, \cdots, z_m]^{\mathrm{T}}$ respectively.

Step 4: The solution vector $x = [x_1, x_2, \cdots, x_{2m+1}]^{\mathrm{T}}$ is given by

$$x_i = \begin{cases} \frac{1}{2}(y_i + z_i) & \text{if } i = 1, 2, \cdots, m, \\ y_{m+1} & \text{if } i = m+1, \\ \frac{1}{2}(y_{2m+2-i} - z_{2m+2-i}) & \text{if } i = m+2, m+3, \cdots, 2m+1. \end{cases}$$

The Algorithm will be referred to as CENTROSYMM-II algorithm. Concerning the computational cost of the CENTROSYMM-II algorithm:

- The time complexity of Step 1 is $\frac{1}{2}n^2 + O(n)$.
- The time complexity of Step 3 depends on the solvers of the two linear systems.
- The time complexity of Step 4 is O (n) (see also [16]).

It may be convenient to finish this section by giving the following result, whose proof will be omitted.

Theorem 4.1

Let $A = (a_{ij})_{i,j=1}^{n}$ be a non-singular centrosymmetric square matrix of order n. Consider the four linear systems of centrosymmetric type:

$$A\boldsymbol{x} = b, \tag{32}$$

$$AJ\boldsymbol{x} = b, \tag{33}$$

$$JA\boldsymbol{x} = b \tag{34}$$

and

$$A^{R}\boldsymbol{x} = b. \tag{35}$$

Then the two linear systems (32) and (35) are equivalent. The same is true for the linear systems (33) and (34). Moreover, if the common solution of the systems (32) and (35) is $x = (k_1, k_2, L, k_n)^T$ then the common solution of the systems (33) and (34) is $x = (k_n, k_{n-1}, L, k_1)^T$.

COMPUTER PROGRAM

In this section, we are going to introduce a MAPLE procedure for solving centrosymmetric linear systems (26) and (29). This procedure is based on the CENTROSYMM-I and CENTROSYMM-II Algorithms.

```
> restart:
centrosymm:=proc(R::array,f::vector,n::posint)
local i, r,m,f1,f2,A,Jm,J,H,y,x,B,Y,Z,X:
global xsoln,detR,detP,detQ,P,Q;
X:= vector(n): m:=floor(n/2): J:=array(1..m, 1..m,
sparse):
for i to m do J[m+1-i,i]:=1 od:
A:= linalg [submatrix](R,1..m,1..m):
if n=2*m then
```

```
#Case (1): n is even #
#Step 1 in CENTROSYMM-I Algorithm.
B:= linalg[submatrix](R,m+1..n,1..m):
P:=evalm( A + evalm(J&*B)):
Q:=evalm( A -evalm(J&*B)):
#Step 2 in CENTROSYMM-I Algorithm.
detP:=linalg[det](P):          detQ:=linalg[det](Q):
detR:=detP *detQ:
if detR = 0 then ERROR("Singular Matrix !!!! ") fi;
#Step 3 in CENTROSYMM-I Algorithm.
f1:=array (1..m): f2:=array(1..m):
for i to m do f1[i]:=f[i]+f[2*m+1-i];
f2[i]:=f[i]-f[2*m+1-i];
od;
Y:=array(1..m):Z:=array(1..m):
Y:=  linalg[linsolve](P,f1):   Z:=linalg[linsolve]
(Q,f2):
#Step 4 in CENTROSYMM-I Algorithm.
for i to m do X[i]:=1/2*(Y[i] +Z[i]);
od;
for i from m+1 to n do X[i]:=1/2*(Y[n+1-i]-Z[n+1-
i]);
od;
xsoln:=simplify([seq(X[r],r=1..n)]):
else
#Case (2): n is odd #
#Step 1 in CENTROSYMM-II Algorithm.
B:= linalg[submatrix](R,m+2..n,1..m):
H:= evalm( A + evalm(J&*B)):
y:=linalg[submatrix](R,1..m,m+1..m+1):
x:=linalg[submatrix](R,m+1..m+1,1..m):
P:=linalg[blockmatrix](2,2,[H,2*y,x,[2]]):op(P);
Q:=evalm( A -evalm(J&*B)):
#Step 2 in CENTROSYMM-II Algorithm.
detP:=linalg[det](P): detQ:=linalg[det](Q): detR:=
detP *detQ:
```

```
if detR = 0 then ERROR("Singular Matrix !!!!") fi;
#Step 3 in CENTROSYMM-II Algorithm.
f1:=array (1..m+1): f2:=array(1..m):
for i to m do f1[i]:=f[i]+f[2*m+2-i];
f2[i]:=f[i]-f[2*m+2-i];
od;
f1[m+1]:=f[m+1];
Y:=array(1..m+1):
Y:=  linalg[linsolve](P,f1):   Z:=linalg[linsolve]
(Q,f2):
#Step 4 in CENTROSYMM-II Algorithm.
for i to m do X[i]:=1/2*(Y[i]+Z[i]);
od;
X[m+1]:=Y[m+1];
for i from m+2 to n do X[i]:=1/2*(Y[n+1-i]-Z[n+1-
i]);
od;
xsoln:=simplify([seq(X[r],r=1..n)]);
fi:
end proc:
```

ILLUSTRATIVE EXAMPLES

All results in this section are obtained with the help of the MAPLE procedure centrosymm.

Example 6.1

Solve the centrosymmetric linear system

$$\begin{bmatrix} 3 & -1 & 1 & 0 & 4 & 1 & -3 & 2 & -3 & 1 \\ 2 & -1 & -3 & 4 & -2 & -1 & -1 & 0 & 3 & 2 \\ 1 & 0 & 2 & -2 & 3 & 3 & 2 & 1 & 3 & 1 \\ 3 & 3 & -3 & 4 & -1 & 4 & -2 & 3 & -1 & 1 \\ 1 & 1 & 2 & -1 & 3 & 2 & 1 & -1 & 3 & 0 \\ 0 & 3 & -1 & 1 & 2 & 3 & -1 & 2 & 1 & 1 \\ 1 & -1 & 3 & -2 & 4 & -1 & 4 & -3 & 3 & 3 \\ 1 & 3 & 1 & 2 & 3 & 3 & -2 & 2 & 0 & 1 \\ 2 & 3 & 0 & -1 & -1 & -2 & 4 & -3 & -1 & 2 \\ 1 & -3 & 2 & -3 & 1 & 4 & 0 & 1 & -1 & 3 \end{bmatrix} \begin{bmatrix} x_1 \\ x_2 \\ x_3 \\ x_4 \\ x_5 \\ x_6 \\ x_7 \\ x_8 \\ x_9 \\ x_{10} \end{bmatrix} = \begin{bmatrix} -1 \\ 11 \\ 9 \\ 16 \\ 6 \\ 9 \\ 4 \\ 7 \\ 2 \\ 6 \end{bmatrix}.$$

Solution:

Here n=10 and m=5. Using the procedure centrosymm, we get the following results: $\hat{b} = [5,13,16,20,15]^T$ and $\tilde{b} = [-7,9,2,12,-3]^T$ $|R| = |P| \times |Q| = -730912$.

$$Q = \begin{bmatrix} 2 & 2 & -1 & 3 & 3 \\ 0 & -4 & -3 & 5 & -1 \\ 0 & -3 & 1 & -4 & 0 \\ 2 & 4 & -6 & 6 & -5 \\ 1 & -2 & 3 & -2 & 1 \end{bmatrix}, \quad P = \begin{bmatrix} 4 & -4 & 3 & -3 & 5 \\ 4 & 2 & -3 & 3 & -3 \\ 2 & 3 & 3 & 0 & 6 \\ 4 & 2 & 0 & 2 & 3 \\ 1 & 4 & 1 & 0 & 5 \end{bmatrix},$$

The solutions of the systems are $y = [2,1,-1,-2,2]^T$ and $z = [0,-1,-1,0,-2]^T$.

$$\begin{bmatrix} 4 & -4 & 3 & -3 & 5 \\ 4 & 2 & -3 & 3 & -3 \\ 2 & 3 & 3 & 0 & 6 \\ 4 & 2 & 0 & 2 & 3 \\ 1 & 4 & 1 & 0 & 5 \end{bmatrix} \begin{bmatrix} y_1 \\ y_2 \\ y_3 \\ y_4 \\ y_5 \end{bmatrix} = \begin{bmatrix} 5 \\ 13 \\ 16 \\ 20 \\ 15 \end{bmatrix}$$

and

$$\begin{bmatrix} 2 & 2 & -1 & 3 & 3 \\ 0 & -4 & -3 & 5 & -1 \\ 0 & -3 & 1 & -4 & 0 \\ 2 & 4 & -6 & 6 & -5 \\ 1 & -2 & 3 & -2 & 1 \end{bmatrix} \begin{bmatrix} z_1 \\ z_2 \\ z_3 \\ z_4 \\ z_5 \end{bmatrix} = \begin{bmatrix} -7 \\ 9 \\ 2 \\ 12 \\ -3 \end{bmatrix}$$

Therefore, we have

$$x_6 = \frac{1}{2}(y_5 - z_5) = 2, \quad x_1 = \frac{1}{2}(y_1 + z_1) = 1,$$

$$x_7 = \frac{1}{2}(y_4 - z_4) = 1, \quad x_2 = \frac{1}{2}(y_2 + z_2) = 0,$$

$$x_8 = \frac{1}{2}(y_3 - z_3) = 0, \quad x_3 = \frac{1}{2}(y_3 + z_3) = -1,$$

$$x_9 = \frac{1}{2}(y_2 - z_2) = 1, \quad x_4 = \frac{1}{2}(y_4 + z_4) = 1,$$

$$x_{10} = \frac{1}{2}(y_1 - z_1) = 1. \quad x_5 = \frac{1}{2}(y_5 + z_5) = 0,$$

Hence the solution vector is $x = [1,0,-1,1,0,2,1,0,1,1]^T$.

Example 6.2

Solve the centrosymmetric linear system

$$\begin{bmatrix} 2 & 1 & 0 & 0 & 0 & 0 & 0 & 0 & 0 & 0 & 0 \\ 1 & 2 & 1 & 0 & 0 & 0 & 0 & 0 & 0 & 0 & 0 \\ 0 & 1 & 2 & 1 & 0 & 0 & 0 & 0 & 0 & 0 & 0 \\ 0 & 0 & 1 & 2 & 1 & 0 & 0 & 0 & 0 & 0 & 0 \\ 0 & 0 & 0 & 1 & 2 & 1 & 0 & 0 & 0 & 0 & 0 \\ 0 & 0 & 0 & 0 & 1 & 2 & 1 & 0 & 0 & 0 & 0 \\ 0 & 0 & 0 & 0 & 0 & 1 & 2 & 1 & 0 & 0 & 0 \\ 0 & 0 & 0 & 0 & 0 & 0 & 1 & 2 & 1 & 0 & 0 \\ 0 & 0 & 0 & 0 & 0 & 0 & 0 & 1 & 2 & 1 & 0 \\ 0 & 0 & 0 & 0 & 0 & 0 & 0 & 0 & 1 & 2 & 1 \\ 0 & 0 & 0 & 0 & 0 & 0 & 0 & 0 & 0 & 1 & 2 \end{bmatrix} \begin{bmatrix} x_1 \\ x_2 \\ x_3 \\ x_4 \\ x_5 \\ x_6 \\ x_7 \\ x_8 \\ x_9 \\ x_{10} \\ x_{11} \end{bmatrix} = \begin{bmatrix} 2 \\ 1 \\ 1 \\ 3 \\ 4 \\ 3 \\ 1 \\ 1 \\ 3 \\ 4 \\ 3 \end{bmatrix}.$$

Solution:

Here n=11 and m=5. Using the procedure centrosymm, we obtain the following results: $\hat{b} = [5,5,4,4,5,3]^T$ and $\tilde{b} = [-1,-3,-2,2,3]^T$ $|R| = |P| \times |Q| = 12$.

$$Q = \begin{bmatrix} 2 & 1 & 0 & 0 & 0 \\ 1 & 2 & 1 & 0 & 0 \\ 0 & 1 & 2 & 1 & 0 \\ 0 & 0 & 1 & 2 & 1 \\ 0 & 0 & 0 & 1 & 2 \end{bmatrix}, \quad P = \begin{bmatrix} 2 & 1 & 0 & 0 & 0 & 0 \\ 1 & 2 & 1 & 0 & 0 & 0 \\ 0 & 1 & 2 & 1 & 0 & 0 \\ 0 & 0 & 1 & 2 & 1 & 0 \\ 0 & 0 & 0 & 1 & 2 & 2 \\ 0 & 0 & 0 & 0 & 1 & 2 \end{bmatrix},$$

The solutions of the systems are $y = [2,1,1,1,1,1]^T$ and $z = [0,-1,-1,1,1]^T$.

$$\begin{bmatrix} 2 & 1 & 0 & 0 & 0 & 0 \\ 1 & 2 & 1 & 0 & 0 & 0 \\ 0 & 1 & 2 & 1 & 0 & 0 \\ 0 & 0 & 1 & 2 & 1 & 0 \\ 0 & 0 & 0 & 1 & 2 & 2 \\ 0 & 0 & 0 & 0 & 1 & 2 \end{bmatrix} \begin{bmatrix} y_1 \\ y_2 \\ y_3 \\ y_4 \\ y_5 \\ y_6 \end{bmatrix} = \begin{bmatrix} 5 \\ 5 \\ 4 \\ 4 \\ 5 \\ 3 \end{bmatrix}$$

and

$$\begin{bmatrix} 2 & 1 & 0 & 0 & 0 \\ 1 & 2 & 1 & 0 & 0 \\ 0 & 1 & 2 & 1 & 0 \\ 0 & 0 & 1 & 2 & 1 \\ 0 & 0 & 0 & 1 & 2 \end{bmatrix} \begin{bmatrix} z_1 \\ z_2 \\ z_3 \\ z_4 \\ z_5 \end{bmatrix} = \begin{bmatrix} -1 \\ -3 \\ -2 \\ 2 \\ 3 \end{bmatrix}$$

Therefore, we get

$$x_7 = \frac{1}{2}(y_5 - z_5) = 0, \quad x_1 = \frac{1}{2}(y_1 + z_1) = 1,$$

$$x_8 = \frac{1}{2}(y_4 - z_4) = 0. \quad x_2 = \frac{1}{2}(y_2 + z_2) = 0,$$

$$x_9 = \frac{1}{2}(y_3 - z_3) = 1. \quad x_3 = \frac{1}{2}(y_3 + z_3) = 0,$$

$$x_{10} = \frac{1}{2}(y_2 - z_2) = 1. \quad x_4 = \frac{1}{2}(y_4 + z_4) = 1,$$

$$x_{11} = \frac{1}{2}(y_1 - z_1) = 1. \quad x_5 = \frac{1}{2}(y_5 + z_5) = 1,$$

$x_6 = y_6 = 1,$

Hence the solution vector is $\mathrm{x} = [1,0,0,1,1,1,0,0,1,1,1]^{\mathrm{T}}$.

Example 6.3

Solve the centrosymmetric linear system

$$\begin{bmatrix} 1 & 0 & -1 & 2 & 3 & -2 & 1 & 0 & 4 & 1 \\ 2 & 1 & 3 & -1 & 6 & 6 & -1 & 3 & 1 & 2 \\ 3 & 2 & 1 & 0 & -1 & 2 & 1 & 1 & 3 & 4 \\ 4 & 3 & 2 & 1 & 0 & 1 & 3 & 4 & 0 & 1 \\ 5 & 4 & 3 & 0 & 1 & 2 & 4 & 5 & 1 & 3 \\ 3 & 1 & 5 & 2 & 2 & 1 & 2 & 3 & 4 & 5 \\ 1 & 0 & 4 & 3 & 1 & 0 & 1 & 2 & 3 & 4 \\ 4 & 3 & 1 & 1 & 2 & -1 & 0 & 1 & 2 & 3 \\ 2 & 1 & 3 & -1 & 6 & 6 & -1 & 3 & 1 & 2 \\ 1 & 4 & 0 & 1 & -2 & 3 & 2 & -1 & 0 & 1 \end{bmatrix} \begin{bmatrix} x_1 \\ x_2 \\ x_3 \\ x_4 \\ x_5 \\ x_6 \\ x_7 \\ x_8 \\ x_9 \\ x_{10} \end{bmatrix} = \begin{bmatrix} -1 \\ 2 \\ 1 \\ 3 \\ 0 \\ 3 \\ 1 \\ 1 \\ 4 \\ 3 \end{bmatrix}.$$

Solution:

Here n=10 and m=5 Using the procedure centrosymm, we obtain the following results:

Error Singular Matrix!!!! This means that the given system has no solutions.

REFERENCES

1. A. L. Andrew, "Centrosymmetric Matrices," SIAM Review, Vol. 40, No. 3, 1998, pp. 697-699. http://dx.doi.org/10.1137/S0036144597328341
2. A. Cantoni and P. Butler, "Eigenvalues and Eigenvectors of Symmetric Centrosymmetric Matrices," Linear Algebra and Its Applications, Vol. 13, No. 3, 1976, pp. 275–288.

3. A. Cantoni and P. Butler, "Properties of the Eigenvectors of Persymmetric Matrices with Applications to Communication Theory," IEEE Transactions on Communications, Vol. 24, No. 8, 1976, pp. 804-809. http://dx.doi.org/10.1109/TCOM.1976.1093391
4. W. Chen, Y. Yu and X. Wang, "Reducing the Computational Requirement of Differential Quadrature Method," Numerical Methods for Partial Differential Equations, Vol. 12, 1996, pp. 565-577.
5. L. Datta and S. Morgera, "Some Results on Matrix Symmetries and a Pattern Recognition Application," IEEE Transactions on Signal Processing, Vol. 34, No. 4, 1986, pp. 992-994.
6. L. Datta and S. Morgera, "On the Reducibility of Centrosymmetric Matrices—Applications in Engineering Problems," Circuits, Systems and Signal Processing, Vol. 8, No. 1, 1989, pp. 71-96.
7. J. Delmas, "On Adaptive EVD Asymptotic Distribution of Centro-Symmetric Covariance Matrices," IEEE Transactions on Signal Processing, Vol. 47, No. 5, 1999, pp. 1402-1406.
8. I. J. Good, "The Inverse of a Centrosymmetric Matrix," Technometrics, Vol. 12, No. 4, 1970, pp. 925-928.
9. Z.-Y. Liu, "Some Properties of Centrosymmetric Matrices," Applied Mathematics and Computation, Vol. 141, No. 2-3, 2003, pp. 297-306.
10. R. B. Mattingly, "Even Order Regular Magic Squares Are Singular," The American Mathematical Monthly, Vol. 107, No. 9, 2000, pp. 777-782.
11. F. Stenger, "Matrices of Sinc Methods," Journal of Computational and Applied Mathematics, Vol. 86, No. 1, 1997, pp. 297-310.
12. J. Weaver, "Centrosymmetric (Cross-Symmetric) Matrices, Their Basic Properties, Eigenvalues, and Eigenvectors," The American Mathematical Monthly, Vol. 92, No. 10, 1985, pp. 711-717.
13. A. C. Aitken, "Determinants and Matrices," Oliver and Boyd, Edinburgh, 1956.
14. H.-T. Gao, C.-H. You and Y. Yang, "An Iterative Method for Generalized Centro-Symmetric Solution of Matrix Equation $AXA^T + BYB^T = C$," 2008. http://www.paper.edu.cn/en_releasepaper/content/21924.
15. A. Graovac, O. Ori, M. Faghani and A. R. Ashrafi, "Distance Property of Fullerenes," Iranian Journal of Mathematical Chemistry, Vol. 2, No. 1, 2011, pp. 99-107.
16. I. T. Abu-Jeib, "Algorithms for Centrosymmetric and Skew-Centrosymmetric Matrices," Missouri Journal of Mathematical Sciences, Vol. 18, No. 1, 2006, pp. 46-53.
17. O. Krafft and M. Schaefer, "Centrogonal Matrices," Linear Algebra and its Applications, Vol. 306, No., 2000, pp. 145-154.
18. Z.-Y. Liu, "Some Properties of Centrosymmetric Matrices and Its Applications," Numerical Mathematics: A Journal of Chinese Universities, Vol. 14, No. 2, 2005, pp. 136-148.
19. Z. Tian and C. Gu, "The Iterative Methods for Centrosymmetric Matrices," Applied Mathematics and Computation, Vol. 187, No. 2, 2007, pp. 902-911.

20. R. Vein and P. Dale, "Determinants and Their Applications in Mathematical Physics," Springer, New York, 1999.
21. A. M. Yasuda, "Some Properties of Commuting and AntiCommuting m-Involutions," Acta Mathematica Scientia, Series B (English Edition), Vol. 32, No. 2, 2012, pp. 631-644.
22. M. E. A. El-Mikkawy, "A Generalized Symbolic Thomas algoriThm," Applied Mathematics, Vol. 3, No. 4, 2012, pp. 342-345. http://dx.doi.org/10.4236/am.2012.34052
23. T. Sugimoto, "On an Inverse Formula of a Tridiagonal Matrix," Operators and Matrices, Vol. 6, No. 3, 2012, pp. 465-480. http://dx.doi.org/10.7153/oam-06-30
24. M. E. A. El-Mikkawy, "A Note on a Three-Term Recurrence for a Tridiagonal Matrix," Journal of Applied Mathematics and Computing, Vol. 139, No. 2-3, 2003, pp. 503-511.
25. M. E. A. El-Mikkawy, "An Algorithm for Solving Tridiagonal Systems," Journal of Institute of Mathematics and Computer Science, Vol. 4, No. 2, 1991, pp. 205-210.
26. M. E. A. El-Mikkawy, "On the Inverse of a General Tridiagonal Matrix," Applied Mathematics and Computation, Vol. 150, No. 3, 2004, pp. 669-679.
27. M. E. A. El-Mikkawy, "A New Computational Algorithm for Solving Periodic Tri-Diagonal Linear Systems," Applied Mathematics and Computation, Vol. 161, No. 2, 2005, pp. 691-696.
28. M. E. A. El-Mikkawy and A. Karawia, "Inversion of General Tridiagonal Matrices," Applied Mathematics Letters, Vol. 19, No. 8, 2006, pp. 712-720.
29. M. E. A. El-Mikkawy and E.-D. Rahmo, "A New Recursive Algorithm for Inverting General Tridiagonal and Anti-Tridiagonal Matrices," Applied Mathematics and Computation, Vol. 204, No. 1, 2008, pp. 368-372.
30. R. L. Burden and J. D. Faires, "Numerical Analysis," 7th ed., Books & Cole Publishing, Pacific Grove, 2001.
31. M. E. A. El-Mikkawy, "A Fast Algorithm for Evaluating nth Order Tri-Diagonal Determinants," Journal of Computational and Applied Mathematics, Vol. 166, No. 2, 2004, pp. 581-584.

CITATION

1. M. El-Mikkawy and F. Atlan, "On Solving Centrosymmetric Linear Systems," Applied Mathematics, Vol. 4 No. 12A, 2013, pp. 21-32. doi: 10.4236/am.2013.412A003.

Numerical Quadratures Using the Interpolation Method of Hurwitz-Radon Matrices

Dariusz Jacek Jakóbczak

Department of Computer Science and Management, Faculty of Electronics and Informatics, Technical University of Koszalin, Koszalin, Poland

5

ABSTRACT

Mathematics and computer sciences need suitable methods for numerical calculations of integrals. Classical methods, based on polynomial interpolation, have many weak sides: they are useless to interpolate the function that fails to be differentiable at one point or differs from the shape of polynomials considerably. We cannot forget about the Runge's phenomenon. To deal with numerical interpolation and integration dedicated methods should be constructed. One of them, called by author the method of Hurwitz-Radon Matrices (MHR) can be used in reconstruction and interpolation of curves in the plane. This novel method is based on a family of Hurwitz-Radon (HR) matrices. The matrices are skew-symmetric and possess columns composed of orthogonal vectors. The operator of Hurwitz-Radon (OHR) built from that matrices, is described. It is shown how to create the orthogonal and discrete OHR and how to use it in a process of function interpolation and numerical integration. Created from the family of N-1 HR matrices and completed with the identical matrix, system of matrices is orthogonal only for vector spaces of dimensions N = 2, 4 or 8. Orthogonality of columns and rows is very significant for stability and high precision of calculations. MHR method is interpolating the curve point by point without using any formula of function. Main features of MHR method are: accuracy of curve reconstruction depending on number of nodes and method of choosing nodes; interpolation of L points of the curve is connected with the computational cost of rank O(L); MHR interpolation is not a linear interpolation.

INTRODUCTION

The following question is important in mathematics and computer sciences: is it possible to find a method of function interpolation and numerical integration without building the interpolation polynomials? Our paper aims at giving the positive answer to this question. This work is the effect of author studies on doctoral thesis. Current methods of numerical calculation of integrals are mainly based on classical polynomial interpolation: Newton, Lagrange or Hermite polynomials and spline curves which are piecewise polynomials [1] [2]. Classical methods are useless to interpolate the function that fails to be differentiable at one point, for example the absolute value function $f(x) = |x|$ at $x = 0$. If point (0; 0) is one of the interpolation nodes, then precise polynomial interpolation of the absolute value function is impossible. Also when the graph of interpolated function differs from the shape of polynomials considerably, for example $f(x) = 1/x$, interpolation is very hard because of existing local extrema of polynomial. We cannot forget about the Runge's phenomenon: when interpolation nodes are equidistance then high-order polynomial oscillates toward the end of the interval, for example close to −1 and 1 with function $f(x) = 1/(1 + 25x^2)$ [3].

This paper deals with the problem of interpolation [4] [5] and numerical quadrature without computing the polynomial or any fixed function. Values of nodes are used to build the orthogonal matrix operators and a linear (convex) combination of OHR operators leads to calculation of curve points. Main idea of MHR method is that the curve is interpolated point by point and computing the unknown coordinates of the points. The only significant factors in MHR method are choosing the interpolation nodes and fixing the dimension of HR matrix (N = 2, 4 or 8). Other characteristic features of function, such as shape or similarity to polynomials, derivative or Runge's phenomenon, are not important in the process of MHR interpolation. The curve is parameterized for value a Î [0;1] in the range of two successive interpolation nodes. In this paper computational algorithm is considered and then we have to talk about time. Complexity of calculations for one unknown point in Lagrange or Newton interpolation based on n nodes is connected with the computational cost of rank O (n^2). Complexity of calculations for L unknown points in MHR interpolation based on n nodes is connected with the computational cost of rank O (L). This is a very important feature of MHR method.

THE INTERPOLATION METHOD OF HURWITZ-RADON MATRICES

Adolf Hurwitz (1859-1919) and Johann Radon (1887-1956) published the papers about specific class of matrices in 1923. Matrices A_i, i=1, 2, L, m satisfying

$$A_j A_k + A_k A_j = 0, \quad A_j^2 = -I \tag{1}$$

For j≠ k; j, k=1, 2, L, m, are called a family of Hurwitz-Radon matrices. A family of HR matrices in Equation (1) has important features: HR matrices are skew-symmetric $\left(A_i^T = -A_i\right)$ and reverse matrix $\left(A_i^{-1} = -A_i\right)$. Only for dimension N = 2, 4 or 8 the family of Hurwitz-Radon matrices consists of N-1 matrices [6].

For N = 2 we have one matrix: $A_1 = \begin{bmatrix} 0 & 1 \\ -1 & 0 \end{bmatrix}$.

For N = 4 there are three matrices with integer entries:

$$A_1 = \begin{bmatrix} 0 & 1 & 0 & 0 \\ -1 & 0 & 0 & 0 \\ 0 & 0 & 0 & -1 \\ 0 & 0 & 1 & 0 \end{bmatrix},$$

$$A_2 = \begin{bmatrix} 0 & 0 & 1 & 0 \\ 0 & 0 & 0 & 1 \\ -1 & 0 & 0 & 0 \\ 0 & -1 & 0 & 0 \end{bmatrix},$$

$$A_3 = \begin{bmatrix} 0 & 0 & 0 & 1 \\ 0 & 0 & -1 & 0 \\ 0 & 1 & 0 & 0 \\ -1 & 0 & 0 & 0 \end{bmatrix}.$$

For N = 8 we have seven matrices with elements 0, ±1 [7].

Let's assume there is given a finite set of points of the function, called further nodes (x_i, y_i) Î R^2 such as:

1. Nodes (interpolation points) are settled at local extrema (maximum or minimum) and at least one point between two successive local extrema;
2. There are five nodes or more.

Assume that the nodes belong to a curve in the plane. How the whole curve could be reconstructed using this discrete set of nodes? Proposed method [8] [9] is based on local, orthogonal matrix operators. Values of nodes' coordinates (x_i, y_i) are connected with HR matrices [10] build on N dimensional vector space. It is important that HR matrices are skew-symmetric and only for dimension N = 2, 4 or 8 columns and rows of HR matrices are orthogonal [1]">]">1].

If one curve is described by a set of nodes $\{(x_i, y_i), i = 1, 2, L, n\}$ then HR matrices combined with identity matrix are used to build an orthogonal and discrete Hurwitz-Radon Operator (OHR) For nodes (x_1, y_1), (x_2, y_2) OHR of dimension N = 2 is constructed:

$$M = \frac{1}{x_1^{\,2} + x_2^{\,2}} \begin{bmatrix} x_1 & x_2 \\ -x_2 & x_1 \end{bmatrix} \begin{bmatrix} y_1 & -y_2 \\ y_2 & y_1 \end{bmatrix} \tag{2}$$

For nodes (x_1, y_1), (x_2, y_2), (x_3, y_3), (x_4, y_4) OHR of dimension N = 4 is constructed:

$$M = \frac{1}{x_1^2 + x_2^2 + x_3^2 + x_4^2} \begin{bmatrix} u_0 & u_1 & u_2 & u_3 \\ -u_1 & u_0 & -u_3 & u_2 \\ -u_2 & u_3 & u_0 & -u_1 \\ -u_3 & -u_2 & u_1 & u_0 \end{bmatrix} \tag{3}$$

Where

$$u_0 = x_1y_1 + x_2y_2 + x_3y_3 + x_4y_4,\ u_1 = -x_1y_2 + x_2y_1 + x_3y_4 - x_4y_3,$$

$$u_2 = -x_1y_3 - x_2y_4 + x_3y_1 + x_4y_2, u_3 = -x_1y_4 + x_2y_3 - x_3y_2 + x_4y_1.$$

For nodes $(x_1, y_1), (x_2, y_2), L, (x_8, y_8)$ OHR M of dimension N = 8 is built [8] similarly as Equation (3):

$$M = \frac{1}{\sum_{i=1}^{8} x_i^2} \begin{bmatrix} u_0 & u_1 & u_2 & u_3 & u_4 & u_5 & u_6 & u_7 \\ -u_1 & u_0 & u_3 & -u_2 & u_5 & -u_4 & -u_7 & u_6 \\ -u_2 & -u_3 & u_0 & u_1 & u_6 & u_7 & -u_4 & -u_5 \\ -u_3 & u_2 & -u_1 & u_0 & u_7 & -u_6 & u_5 & -u_4 \\ -u_4 & -u_5 & -u_6 & -u_7 & u_0 & u_1 & u_2 & u_3 \\ -u_5 & u_4 & -u_7 & u_6 & -u_1 & u_0 & -u_3 & u_2 \\ -u_6 & u_7 & u_4 & -u_5 & -u_2 & u_3 & u_0 & -u_1 \\ -u_7 & -u_6 & u_5 & u_4 & -u_3 & -u_2 & u_1 & u_0 \end{bmatrix} \tag{4}$$

Where

$$u=\begin{bmatrix} y_1 & y_2 & y_3 & y_4 & y_5 & y_6 & y_7 & y_8 \\ -y_2 & y_1 & -y_4 & y_3 & -y_6 & y_5 & y_8 & -y_7 \\ -y_3 & y_4 & y_1 & -y_2 & -y_7 & -y_8 & y_5 & y_6 \\ -y_4 & -y_3 & y_2 & y_1 & -y_8 & y_7 & -y_6 & y_5 \\ -y_5 & y_6 & y_7 & y_8 & y_1 & -y_2 & -y_3 & -y_4 \\ -y_6 & -y_5 & y_8 & -y_7 & y_2 & y_1 & y_4 & -y_3 \\ -y_7 & -y_8 & -y_5 & y_6 & y_3 & -y_4 & y_1 & y_2 \\ -y_8 & y_7 & -y_6 & -y_5 & y_4 & y_3 & -y_2 & y_1 \end{bmatrix} \cdot \begin{bmatrix} x_1 \\ x_2 \\ x_3 \\ x_4 \\ x_5 \\ x_6 \\ x_7 \\ x_8 \end{bmatrix} \tag{5}$$

The components of the vector $u=(u_0,u_1,L,u_7,)^T$, appearing in the matrix M in Equation (4), are defined by Equation (5) in the similar way to Equations (2) and (3) but in terms of the coordinates of the above 8 nodes. Note that OHR operators in Equations (2)-(4) satisfy the condition of interpolation

$$M \cdot x = y \tag{6}$$

For $x = (X_1, X_2, L, X_N) \in R^N$ x^1 0, $y = (y_1, y_2, L, y_N) \in R^N$, N = 2, 4 or 8.

How can we compute coordinates of points settled between interpolation nodes? On a segment of a line every number "c" situated between "a" and "b" is described by a linear (convex) combination $c = \alpha \cdot a + (1-\alpha)\, b$ for

$$\alpha = \frac{b-c}{b-a} \in [0;1] \tag{7}$$

Average OHR operator M_2 of dimension N = 2, 4 or 8 is constructed as follows:

$$M_2 = \alpha \cdot M_0 + (1-\alpha) \cdot M_1 \tag{8}$$

With the operator M_0 built in Equations (2)-(4) by "odd" nodes $(x_1 = a, y_1), (x_3, y_3), L, (x_{2N-1}, y_{2N-1})$ and M_1 built in Equations (2)-(4) by "even" nodes $(x_2 = b, y_2), (x_4, y_4), L, (x_{2N}, y_{2N})$.

Notice that having the operator M_2 for coordinates $x_i < x_{i+1}$ it is possible to reconstruct the second coordinates of points (x, y) in terms of the vector C defined with

$$c_i = \alpha \cdot x_{2i-1} + (1-\alpha) \cdot x_{2i} \tag{9}$$

For i=1, 2, L, N. as $C = [c_1, c_2, L, c_N]^T$.

The required formula is adequate to Equation (6):

$$Y(C) = M_2 \cdot C \tag{10}$$

in which components of vector Y(C) give us the second coordinate of the points (x, y) corresponding to the first coordinate, given in terms of components in Equation (9) of the vector C.

After computing in Equations (7)-(10) for any a Î [0;]»>]»>1], we have a half of reconstructed points (j = 1 in Algorithm 1). Now it is necessary to find second half of unknown coordinates (j = 2 in Algorithm 1) for i=1, 2, L, N.

$$c_i = \alpha \cdot x_{2i} + (1-\alpha) \cdot x_{2i+1}, \tag{11}$$

There is no need to build the OHR for nodes $(x_2 = a, y_2), (x_4, y_4), L, (x_{2N}, y_{2N})$ because we just find M_1. This operator will play a role as M_0 in Equation (8). New M_1 must be computed for nodes $(x_3 = b, y_3), L, (x_{2N-1}, y_{2N-1}), (x_{2N+1}, y_{2N+1})$.

As we see the minimum number of interpolation nodes n = 2N + 1 = 5, 9 or 17 using OHR operators of dimension N = 2, 4 or 8 respectively. If there is more nodes than 2N + 1, the same calculations in Equations (7)-(11) have to be done for next range(s) or last range of 2N + 1 nodes. For example, if n = 9 then we can use OHR operators of dimension N = 4 or OHR operators of dimension N = 2 for two subsets of nodes:

$\{(x_1, y_1), L, (x_5, y_5)\}$ and $\{(x_5, y_5), L, (x_9, y_9)\}$. We summarize this section in the following algorithm of point's reconstruction for 2N+1 = 5, 9 or 17 successive nodes.

Algorithm 1: let j = 1.

Input: Set of interpolation nodes {(x_i, y_i), i = 1, 2, ×××, n; n = 5, 9 or 17}.
Step 1. Determine the dimension N of OHR operators: N = 2 if n = 5, N = 4 if n = 9, N = 8 if n = 17.
Step 2. Build M_0 for nodes (x_1 = a, y_1), (x_3, y_3), ×××, (x_{2N-1}, y_{2N-1}) and M_1 for nodes (x_2 = b, y_2), (x_4, y_4), ×××, (x_{2N}, y_{2N}) from Equations (2)-(4).
Step 3. Determine the number of points to be reconstructed $K_j > 0$ between two successive nodes, let k = 1.
Step 4. Compute a Î [0;]»>]»>1] from Equation (7) for c_1 = c = a×a + (1 – a)×b.
Step 5. Build M_2 from Equation (8).
Step 6. Compute vector C = [c_1, c_2, ×××, c_N]^T from Equation (9).
Step 7. Compute unknown coordinates Y(C) from Equation (10).
Step 8. If $k < K_j$, set k = k + 1 and go to Step 4. Otherwise if j = 1, set $M_0 = M_1$, a = x_2, b = x_3, build new M_1 for nodes (x_3, y_3), (x_5, y_5), ×××, (x_{2N+1}, y_{2N+1}), let j = 2 and go to Step 3. Otherwise, stop.

The number of reconstructed points in Algorithm 1 is K = N (K_1 + K_2). If there is more nodes than 2N + 1 = 5, 9 or 17, Algorithm 1 has to be done for next range(s) or last range of 2N + 1 nodes. Reconstruction of curve points using Algorithm 1 is called by author the method of Hurwitz-Radon Matrices (MHR).

MHR CALCULATIONS AND GRAPHICAL EXAMPLES

In this section we consider the number of multiplications and divisions for MHR method during reconstruction of K = L – n points having n interpolation nodes of the curve consists of L points. First we present a formula for computing one unknown coordinate of a single point. Assume there are given four nodes (x_1, y_1), (x_2, y_2), (x_3, y_3) and (x_4, y_4). OHR operators of dimension N = 2 are built in Equation (2) as follows:

$$M_0 = \frac{1}{x_1^2 + x_3^2}\begin{bmatrix} x_1 y_1 + x_3 y_3 & x_3 y_1 - x_1 y_3 \\ x_1 y_3 - x_3 y_1 & x_1 y_1 + x_3 y_3 \end{bmatrix}$$

$$M_1 = \frac{1}{x_2^2 + x_4^2} \begin{bmatrix} x_2 y_2 + x_4 y_4 & x_4 y_2 - x_2 y_4 \\ x_2 y_4 - x_4 y_2 & x_2 y_2 + x_4 y_4 \end{bmatrix}.$$

Let first coordinate c_1 of reconstructed point is situated between x_1 and x_2:

$$c_1 = \alpha \cdot x_1 + \beta \cdot x_2 \quad \text{for} \ \ 0 \le \beta = 1 - \alpha \le 1. \tag{12}$$

Compute second coordinate of reconstructed point $y(c_1)$ for $Y(C) = [y(c_1), y(c_2)]^T$ from Equation (10):

$$\begin{bmatrix} y(c_1) \\ y(c_2) \end{bmatrix} = (\alpha \cdot M_0 + \beta \cdot M_1) \cdot \begin{bmatrix} \alpha \cdot x_1 + \beta \cdot x_2 \\ \alpha \cdot x_3 + \beta \cdot x_4 \end{bmatrix} \tag{13}$$

After calculation by Equation (13):

$$y(c_1) = \alpha^2 \cdot y_1 + \beta^2 \cdot y_2 + \frac{\alpha \cdot \beta}{x_1^2 + x_3^2}(x_1 x_2 y_1 + x_2 x_3 y_3 + x_3 x_4 y_1 - x_1 x_4 y_3)$$
$$+ \frac{\alpha \cdot \beta}{x_2^2 + x_4^2}(x_1 x_2 y_2 + x_1 x_4 y_4 + x_3 x_4 y_2 - x_2 x_3 y_4). \tag{14}$$

So each point of the curve $P = (c_1, y(c_1))$ settled between nodes (x_1, y_1) and (x_2, y_2) is parameterized by P (a) for Equations (12) and (14) and a Î [0; 1].

If nodes (x_i, y_i) are equidistance in coordinate x_i, then parameterization of unknown coordinate (14) is simpler. Let four successive nodes (x_1, y_1), (x_2, y_2), (x_3, y_3) and (x_4, y_4) are equidistance in coordinate x_i and $a = x_1$, $h/2 = x_{i+1} - x_i =$ const. Calculations in Equations (13) and (14) are done for c_1 in Equation (12):

$$y(c_1) = \alpha y_1 + \beta y_2 + \alpha\beta s \tag{15}$$

and

$$s = h\left(\frac{2ay_1 + hy_1 + hy_3}{4a^2 + 4ah + 2h^2} - \frac{2ay_2 + 2hy_2 + hy_4}{4a^2 + 8ah + 5h^2}\right) \tag{16}$$

As we can see in Equations (15) and (16), MHR interpolation is not a linear interpolation. It is possible to estimate the in terpolation error of MHR method (Algorithm 1) for the class of linear function f:

$$\left|f(c_1)-y(c_1)\right| = \left|\alpha y_1 + \beta y_2 - y(c_1)\right| = \alpha\beta\left|s\right| \qquad (17)$$

Notice that estimation from Equation (17) has the biggest value 0.25çsç for b = a = 0.5, when c_1 is situated in the middle between x_1 and x_2.

The goal of this paper is not a reconstruction of single point, like for example Equations (14) and (15), but interpolation of curve consists of L points. If we have n interpolation nodes, then there is K = L – n points to find using Algorithm 1 and MHR method. Now we consider the complexity of MHR calculations.

Lemma 1: Let n = 5, 9 or 17 is the number of interpolation nodes, let MHR method (Algorithm 1) is done for reconstruction of the curve consists of L points. Then MHR method is connected with the computational cost of rank O (L).

Proof: Using Algorithm 1 we have to reconstruct K = L – n points of unknown curve. Counting the number of multiplications and divisions D in Algorithm 1 here are the results:

D = 4L + 7 for n = 5 and L = 2i + 5;
D = 6L + 21 for n = 9 and L = 4i + 9;
D = 10L + 73 for n = 17 and L = 8i + 17; i=2, 3, 4, L.

The lowest computational cost appears in MHR method with five nodes and OHR operators of dimension N = 2. Therefore whole set of n nodes can be divided into subsets of five nodes. Then whole curve is to be reconstructed by Algorithm 1 with all subsets of five nodes:

$\{(x_1,y_1),L,(x_5,y_5)\}$, $\{(x_5,y_5),L,(x_9,y_9)\}$, $\{(x_9,y_9),L,(x_{13},y_{13})\}$, L If the last node (x_n, y_n) is indexed n ≠ 4i + 1 then we have to use lastive nodes $\{(x_{n-4},y_{n-4}),L,(x_n,y_n)\}$ in Algorithm 1.

Function f(x) = 1/x is an example when the graph of interpolated function differs from the shape of polynomials considerably. Then classical interpolation is very hard because of existing local extrema of polynomial (Figure 1). Here is the application of Algorithm 1 for this function and five nodes.

Figure 2 contains not many interpolated points (twenty six) and minimal number of nodes (five), so numerical calculations of integral (precise value I = 2.196) by trapezoidal rule I_1 = 2.213 are not always satisfying. Greater number of nodes and interpolated points gives us more accurate value of quadrature.

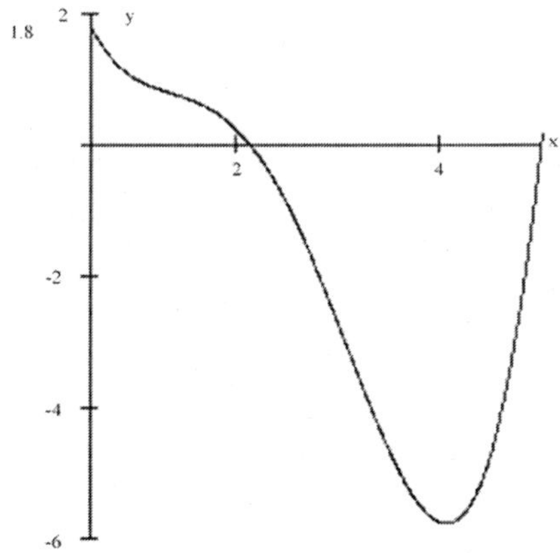

Figure 1: Lagrange interpolation polynomial of function f = 1/x differs extremely from the original.

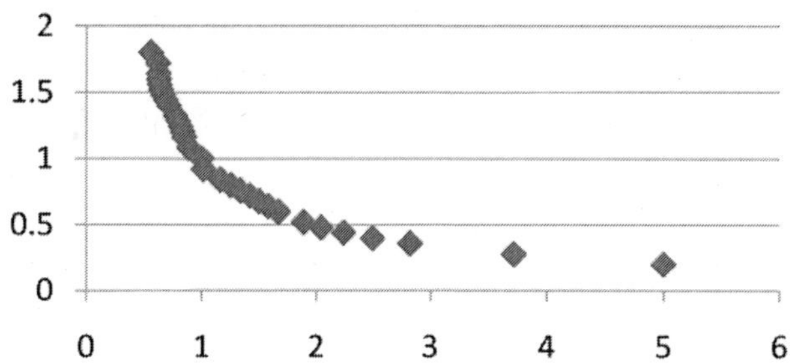

Figure 2: Twenty six interpolated points of function f = 1/x.

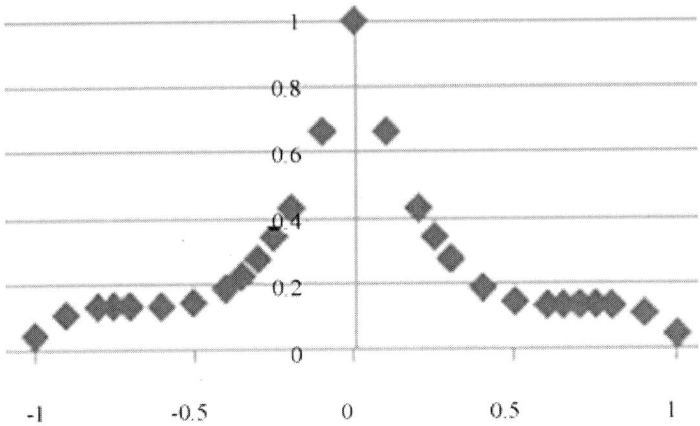

Figure 3: Function f = 1/ (1 + 25x²), interpolated via MHR, without Runge phenomenon.

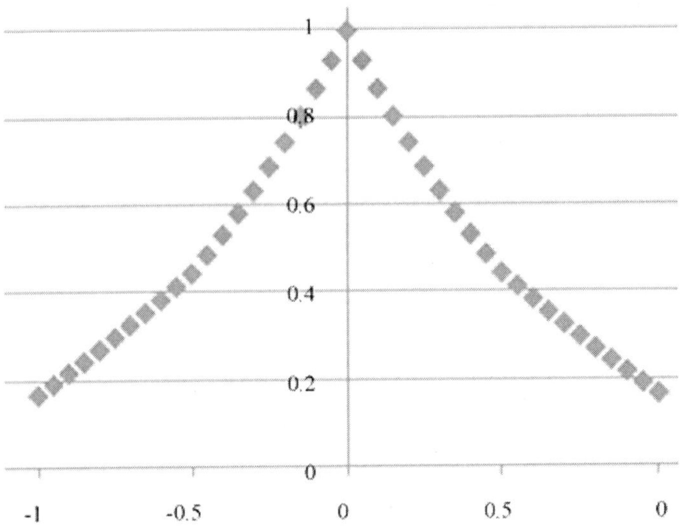

Figure 4: Function f = 1/ (1 + 5x²), interpolated via MHR, without Runge phenomenon.

Lagrange interpolation polynomial for function f(x) = 1/x and nodes (5; 0.2), (5/3; 0.6), (1; 1), (5/7; 1.4), (5/9; 1.8) has one local minimum and two roots.

Other examples of MHR interpolation and numerical quadratures:

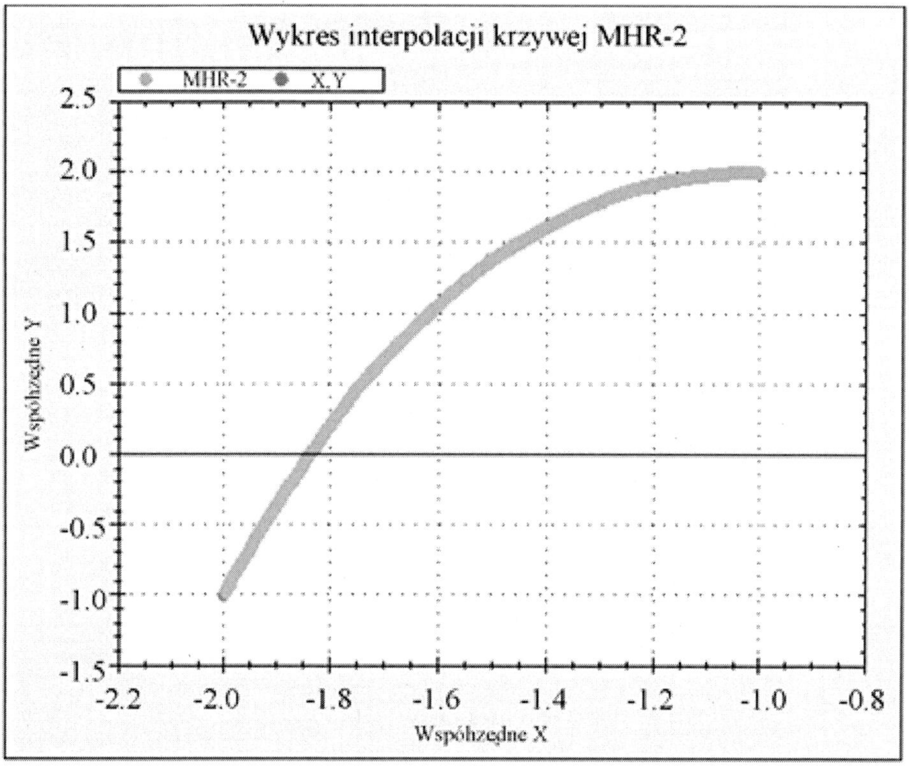

Figure 5: Function $f(x) = x^3 + x^2 - x + 1$ with 396 interpolated points using MHR method with 5 nodes: (–2; –1), (–1.75; 0.453125), (–1.5; 1.375), (–1.25; 1.859375) and (–1; 2).

Figure 3 contains minimal number of nodes (five) and not many interpolated points (twenty two), so numerical calculations of integral (precise value I = 0.549) are not always satisfying:

1. trapezoid method: $I_1 = 0.534$;
2. Simpson's rule: $I_2 = 0.538$.

Greater number of nodes and interpolated points gives us more accurate value of quadrature.

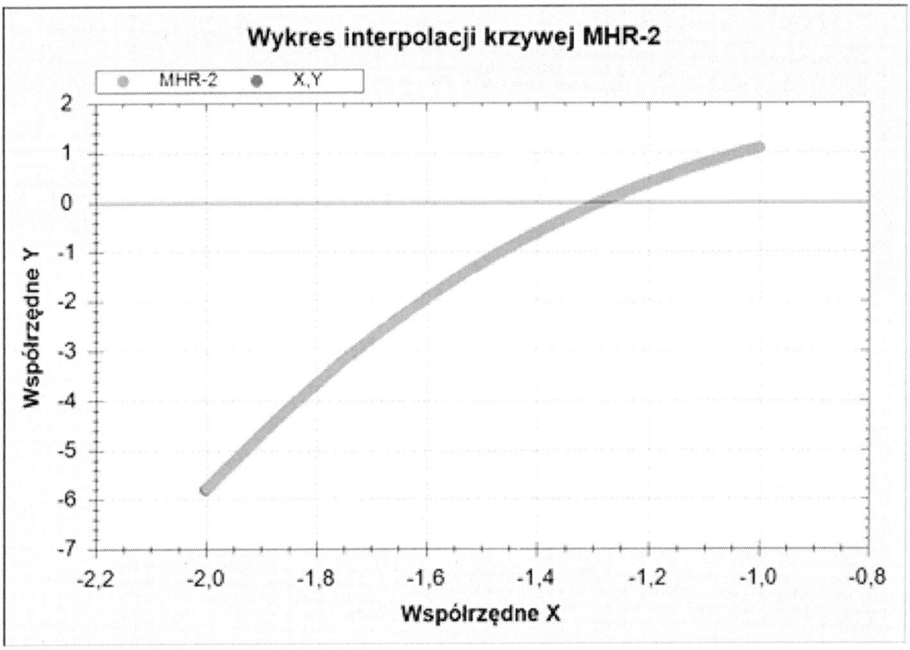

Figure 6: Function $f(x) = x^3 + \ln(7 - x)$ with 396 interpolated points using MHR method with 5 nodes: (–2; –5.803), (–1.75; –3.190), (–1.5; –1.235), (–1.25; 0.1571) and (–1; 1.0794).

Figure 4 contains minimal number of nodes (five) and not many interpolated points (thirty six), but numerical calculations of integral (precise value I = 1.029) are interesting:

1. trapezoid method: $I_1 = 1.000$;
2. Simpson's rule: $I_2 = 0.999$.

After computing of K points for interpolated function (algorithm 1), it is possible to calculate the integral by trapezoid method or Simpson's rule in range of each pair of successive interpolated points. Greater number of nodes and interpolated points gives us more accurate value of quadrature.

Figures 5-8 are the examples of interpolation for such functions as polynomial, logarithmic and exponential.

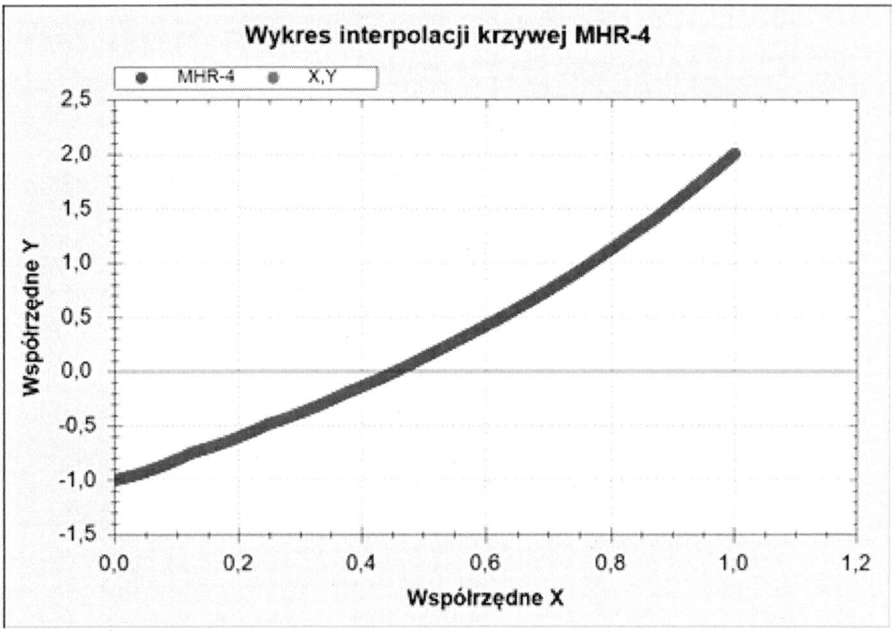

Figure 7: Function $f(x) = x^3 + 2x - 1$ with interpolated points using MHR method with 9 nodes for $x_i = 0; 0.125; 0.25; 0.375; 0.5; 0.625; 0.75; 0.875$ and 1.

SUMMARY

The method of Hurwitz-Radon Matrices (MHR-Algorithm 1) leads to curve interpolation depending on the number of nodes and location of nodes. No characteristic features of curve are important in MHR method: failing to be differentiable at any point, the Runge's phenomenon or differences from the shape of polynomials.

These features are very significant for classical polynomial interpolations. MHR method gives the possibility of curve reconstruction and then numerical calculations of integrals for interpolated function via trapezoid method or Simpson's rule for example. The only condition is to have a set of nodes according to assumptions in Algorithm 1. Curve modeling [12] by MHR method is connected with possibility of changing the nodes coordinates and reconstruction of new curve or contour

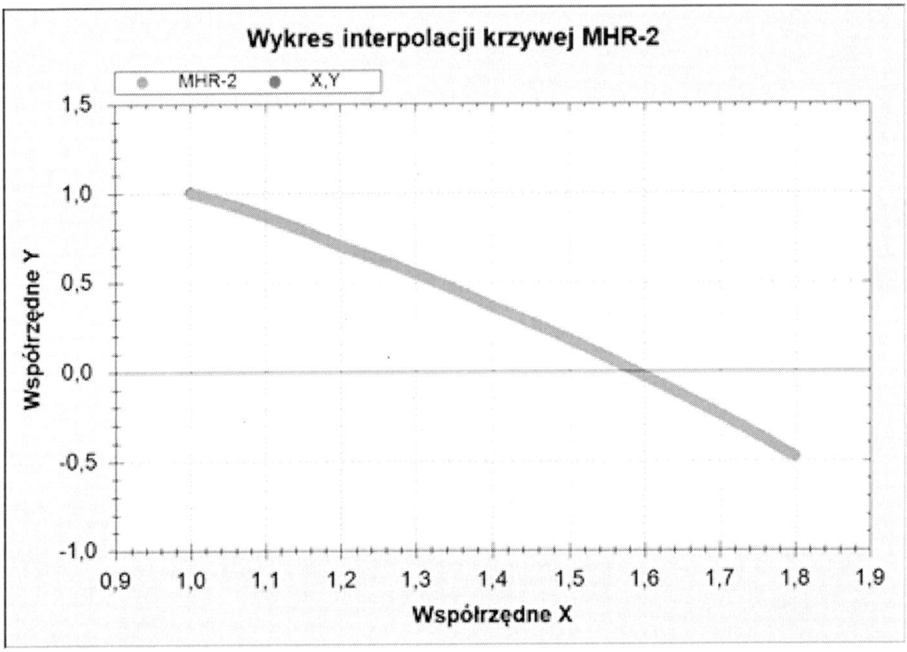

Figure 8: Function $f(x) = 3 - 2^x$ with 396 interpolated points using MHR method with 5 nodes: (1; 1), (1.2; 0.7026), (1.4; 0.361), (1.6; –0.031) and (1.8; –0.482).

for new set of nodes, no matter what shape of curve or contour is to be reconstructed. Main features of MHR method are:

- accuracy of curve reconstruction depending on number of nodes and method of choosing nodes;
- Reconstruction of curve consisting of L points is connected with the computational cost of rank O(L);
- Algorithm 1 is dealing with local operators: average OHR operator M_2 (8) is built by successive 4, 8 or 16 nodes, what is connected with smaller computational costs then uses all nodes.

Future works are connected with: geometrical transformations of curve (translations, rotations, scaling) — only nodes are transformed and new curve (for example contour of the object) for new nodes is reconstructed; estimation of curve length [13]; possibility to apply MHR method to three-dimensional curves [8] [9]; object recognition [14], shape representation [15] [16] and parameterization [17] in image processing; curve extrapolation [18].

REFERENCES

1. Dahlquist, G. and Bjoerck, A. (1974) Numerical Methods. Englewood Cliffs N. J., Prentice-Hall.
2. Jankowska, J. and Jankowski, M. (1981) Survey of Numerical Methods and Algorithms (Part I). Wydawnictwa Naukowo-Techniczne, Warsaw. (in Polish)
3. Ralston, A. (1965) A First Course in Numerical Analysis. McGraw-Hill Book Company, Boston.
4. Kozera, R. (2004) Curve Modeling via Interpolation Based on Multidimensional Reduced Data. Silesian University of Technology Press, Gliwice.
5. Jakóbczak, D. (2009) Curve Interpolation Using Hurwitz-Radon Matrices. Polish Journal of Environmental Studies, 3B, 126-130.
6. Citko, W., Jakóbczak, D. and Sieńko, W. (2005) On Hurwitz-Radon Matrices Based Signal Processing. Signal Processing'2005, Poznań, 19-24.
7. Sieńko, W., Citko, W. and Jakóbczak, D. (2004) Learning and System Modeling via Hamiltonian Neural Networks. Lecture Notes on Artificial Intelligence, 3070, 266-271.
8. Jakóbczak, D. (2007) 2D and 3D Image Modeling Using Hurwitz-Radon Matrices. Polish Journal of Environmental Studies, 16, 104-107.
9. Jakóbczak, D. and Kosiński, W. (2007) Hurwitz-Radon Operator in Monochromatic Medical Image Reconstruction. Journal of Medical Informatics & Technologies, 11, 69-78.
10. Eckmann, B. (1999) Topology, Algebra, Analysis—Relations and Missing Links. Notices of AMS, 46, 520-527.
11. Lang, S. (1970) Algebra. Addison-Wesley Publishing Company, Boston.
12. Jakóbczak, D. (2010) Object Modeling Using Method of Hurwitz-Radon Matrices of Rank k. In: Wolski, W. and Borawski, M., Eds., Computer Graphics: Selected Issues, University of Szczecin Press, Szczecin, 79-90.
13. Jakóbczak, D. (2010) Shape Representation and Shape Coefficients via Method of Hurwitz-Radon Matrices. Lecture Notes in Computer Science, 6374, 411-419.
14. Jakóbczak, D. (2011) Object Recognition via Contour Points Reconstruction Using Hurwitz-Radon Matrices. In: Józefczyk, J. and Orski, D., eds., Knowledge-Based Intelligent System Advancements: Systemic and Cybernetic Approaches, IGI Global, Hershey, 87-107.

15. Jakóbczak, D. (2010) Application of Hurwitz-Radon Matrices in Shape Representation. In: Banaszak, Z. and swic, A. Eds., Applied Computer Science: Modelling of Production Processes, 1, Lublin University of Technology Press, Lublin, 63-74.
16. Jakóbczak, D. (2010) Implementation of Hurwitz-Radon Matrices in Shape Representation. In: Choras, R.S., Ed., Advances in Intelligent and Soft Computing 84, Image Processing and Communications: Challenges 2, Springer-Verlag, Berlin Heidelberg, 39-50.
17. Jakóbczak, D. (2011) Curve Parameterization and Curvature via Method of Hurwitz-Radon Matrices. Image Processing & Communications—An International Journal, 1-2, 49-56.
18. Jakóbczak, D. (2011) Data Extrapolation and Decision Making via Method of Hurwitz-Radon Matrices. Lecture Notes in Computer Science/LNAI, 6922, 173-182.

CITATION

1. Jacek Jakóbczak, D. (2014) Numerical Quadratures Using the Interpolation Method of Hurwitz-Radon Matrices.Advances in Linear Algebra & Matrix Theory, 4, 100-108. doi: 10.4236/alamt.2014.42008.

A Modified Precondition in the Gauss-Seidel Method

Alimohammad Nazari and
Sajjad Zia Borujeni
Department of Mathematics, Faculty of Science, Arak University, Arak, Iran

6

ABSTRACT

In recent years, a number of preconditioners have been applied to solve the linear systems with Gauss-Seidel method (see [1-13]). In this paper we use S_l instead of $(S + S_m)$ and compare with M. Morimoto's precondition [3] and H. Niki's precondition [5] to obtain better convergence rate. A numerical example is given which shows the preference of our method.

INTRODUCTION

Consider the linear system

$$Ax = b, \tag{1}$$

where $A = (\alpha_{ij}) \in R^{n \times n}$ is a known nonsingular matrix and $x, b \in R^n$ are vectors. For any splitting A = M – N with a nonsingular matrix M, the basic splitting iterative method can be expressed as

$$x^{(k+1)} = M^{-1}Nx^{(k)} + M^{-1}b. \qquad k = 0,1,2,\cdots \tag{2}$$

Assume that

$a_{ii} \neq 0, i = 1, 2, \cdots, n,$

without loss of generality we can write

$$A = I - L - U, \tag{3}$$

where I is the identity matrix, -L and -U are strictly lower triangular and strictly upper triangular parts of A, respectively. In order to accelerate the convergence of the iterative method for solving the linear system (1), the original linear system (1) is transformed into the following preconditioned linear system

$$PAx = Pb, \tag{4}$$

where P, called a preconditioner, is a nonsingular matrix.

In 1991, Gunawardena et al. [2] considered the modified Gauss-Seidel method with $P = (I + S)$, where

$$S = (s_{ij}) = \begin{cases} -a_{ii+1} & \text{for } i = 1, 2, \cdots, n-1,\ j = i+1 \\ 0 & \text{otherwise.} \end{cases}$$

Then, the preconditioned matrix $A_S = (I + S)A$ can be written as

$$\begin{aligned} A_S &= I - L - SL - U + S - SU \\ &= (I - D) - (L + E) - (U - S + SU), \end{aligned}$$

where D and E are the diagonal and strictly lower triangular parts of SL, respectively. If $a_{ii+1}a_{i+1i} \neq 1 (1 \leq i \leq n-1)$, then $\{(I - D) - (L + E)\}^{-1}$ exists. Therefore, the preconditioned Gauss-Seidel iterative matrix T_S for A_S becomes

$$T_S = \{(I - D) - (L + E)\}^{-1} (U - S + SU),$$

which is referred to as the modified Gauss-Seidel iterative matrix. Gunawardena et al. proved the following inequality:

$$\rho(T_S) \le \rho(T) < 1,$$

where ρ(T) denotes the spectral radius of the GaussSeidel iterative matrix T. Morimoto et al. [3] have proposed the following preconditioner,

$$P_{s+m} = \left(a_{ij}^{(s+m)}\right) = I + S + S_m.$$

In this preconditioner, S_m is defined by

$$S_m = \left(s_{ij}^{(m)}\right) = \begin{cases} -a_{il_i} & 1 \le i < n-1, i+1 < j \le n, \\ 0 & \text{otherwise,} \end{cases}$$

where $l_i = \min l_i$ and

$l_i = \{j : |a_{ij}| \textit{ is maximal for } i+1 < j \le n\}$, for

$1 \le i < n-1$. The preconditioned matrix $A_{s+m} = (I + S + S_m)A$ can then be written as

$$\begin{aligned} A_{s+m} &= I - L - SL - U + S \\ &\quad - SU + S_m - S_m L - S_m U \\ &= \{(I - D_{s+m}) - (L + E_{s+m})\} \\ &\quad - (U - S - S_m + SU + S_m U + F_{s+m}), \end{aligned}$$

where D_{s+m}, E_{s+m} and F_{s+m} are the diagonal, strictly lower and strictly upper triangular parts of $(S+S_m)L$ respectively. Assume that the following inequalities are satisfied:

$$\begin{cases} 0 < a_{ii+1}a_{i+1i} + a_{il_i}a_{l_ii} < 1 & 1 \le i < n-1 \\ 0 < a_{ii+1}a_{i+1i} < 1 & i = n-1. \end{cases}$$

Then $\{(I-D_{s+m})-(L+E_{s+m})\}$ is nonsingular. The preconditioned Gauss-Seidel iterative matrix T_{s+m} for A_{s+m} is then defined by

$$T_{s+m} = \{(I-D_{s+m})-(L+E_{s+m})\}^{-1} \cdot (U - S - S_m + SU + S_mU + F_{s+m}).$$

Morimoto et al. [3] proved that $\rho(T_{s+m}) \le \rho(T_s)$. To extend the preconditioning effect to the last row, Morimoto et al. [7] proposed the preconditioner

$$P_R = I + R,$$

where R is defined by

$$R = (r_{nj}) = \begin{cases} -a_{nj} & 1 \le j \le n-1 \\ 0 & \text{otherwise} \end{cases}$$

The elements a_{nj}^R of A_R are given by

$$A_R = (I+R)A = \left(a_{ij}^R\right),$$

$$a_{ij}^R = \begin{cases} a_{ij} & 1 \le i < n, 1 \le j \le n \\ a_{nj} - \sum_{k=1}^{n-1} a_{nk}a_{kj} & 1 \le j \le n. \end{cases}$$

And Morimoto et al. proved that $\rho(T_R) \le \rho(T)$ holds, where T_R is the iterative matrix for A_R. They also presented combined preconditioners, which are given by combinations of R with any upper preconditioner, and showed that the convergence rate of the combined methods are better than those of the Gauss-Seidel method applied with other upper preconditioners [7]. In [14], Niki et al. considered the preconditioner $P_{SR} = (I+S+R)$. Denote $A_{SR} = M_{SR} - N_{SR}$. In [5], Niki et al. proved that if the following inequality is satisfied,

$$a_{nj} \le a_{nj}^R = a_{nj} - \sum_{k=1,\, k \ne j}^{n-1} a_{nk}a_{kj} \quad (j = 1, 2, \cdots, n-1), \tag{5}$$

then $\rho(T_{SR}) \le \rho(T_s)$ holds, where T_{SR} is the iterative matrix for A_{SR}. For matrices that do not satisfy Equation

(5), by putting $R = (r_{nj}) = -\sum_{k=1,k\ne j}^{n-1} a_{nk}a_{kj} - a_{nj}, 1 \le j < n,$

Equation (5) is satisfied. Therefore, Niki et al. [5] proposed a new preconditioner $P_G = I + \gamma G (\gamma \ge 1)$, where

$$G = \left(g_{nj}\right) = \begin{cases} \sum_{k=1,\, k\ne j}^{n-1} a_{nk}a_{kj} - a_{nj}, & 1 \le j \le n-1, \\ 0 & \text{otherwise.} \end{cases}$$

Put $A_G = P_G A = (a_{ij}^G) = M_G - N_G$, and $T_G = M_G^{-1} N_G$.

Replacing P_G by $P_G = I + S + S_m + \gamma G$, and setting $\gamma = 1$ the Gauss-Seidel splitting of A_G can be written as

$$A_G = \{(I - D_{s+m}) - (L + E_{s+m}) - (G(L+U) - G)\}$$
$$- (U - S - S_m + SU + S_m U + F_{s+m}),$$

where $G(L+U)-G$ is constructed by the elements $a_{nj}^G = g_{nj} a_{jn}$. Thus, if the preconditioner P_G is used, then all of the rows of A are subject to preconditioning. Niki et al. [5] proved that under the condition $\gamma_u > \gamma$, $\rho(T_R) \geq \rho(T_G)$, where γ_u is the upper bound of those values of γ for which $\rho(T_G) < 1$. By setting $a_{ij}^G = 0$, they obtained

$$\gamma = -a_{nj} \Bigg/ \left(g_{nj} + \sum_{k=1, k \neq j}^{n-1} g_{nk} a_{kj} \right).$$

Niki et al. [5] proved that the preconditioner P_G satisfies the Equation (5) unconditionally. Moreover, they reported that the convergence rate of the GaussSeidel method using preconditioner P_G is better than that of the SOR method using the optimum ω found by numerical computation. They also reported that there is an optimum $\gamma(\gamma_{opt})$ in the range $\gamma_u > \gamma_{opt} > \gamma_m$, which produces an extremely small $\rho(T_{\gamma opt})$, where is the upper bound of the values of γ for which $a_{nj}^G \geq 0$, for all j.

In this paper we use different preconditions for solving (1) by Gauss-Siedel method, that assuming none of the components of the matrix A to be zero. If the largest component of the column j is not $a_{i,i+1},$ then the value of ρ (T) will be improved.

MAIN RESULT

In this section we replace S_l by $(S+S_m)$ of Morimoto such that $S_l = S_n + S_m$ and define S_n by

$$S_n = (s_{ij}) = \begin{cases} -a_{k_i j} & i = 1,2,\dots,n-2,\ j = i+1, \\ -a_{ij} & i = n-1, \\ 0 & \text{otherwise}, \end{cases}$$

where $-a_{k_i j} = \max |a_{kj}|$ s.t $k = i, i+2, i+3, \cdots, n$, and S_m has the same form as the S_m proposed by Morimoto et al. [3].

The precondition Matrix $A_{s_l} = (I + S_l)A$ can then be written as

$$\begin{aligned} A_{S_l} &= I - L - U + S_l - S_l L - S_l U \\ &= \left(I - D_{S_l}\right) - \left(L + E_{S_l}\right) - \left(U - S_l + S_l U + F_{S_l}\right), \end{aligned}$$

where D_{s_l}, E_{s_l} and F_{s_l} are the diagonal, strictly lower and strictly upper triangular parts of $S_l L$, respectively. Assume that the following inequalities are satisfied:

$$\begin{cases} -a_{ii+1} + a_{k_i i+1} + a_{il_i} a_{l_i i+1} \geq 0 & i = 1,2,\cdots,n-2 \\ 0 < a_{k_i i+1} a_{i+1 i} + a_{il_i} a_{l_i i} < 1 & i = 1,2,\cdots,n-2 \\ 0 < a_{ii+1} a_{i+1 i} < 1 & i = n-1, \end{cases} \tag{6}$$

Therefore $M_{s_l}^{-1}$ exists and the preconditioned GaussSeidel iterative matrix T_{s_l} for A_{s_l} is defined by

$$\begin{aligned}T_{S_l} &= M_{S_l}^{-1} N_{S_l} \\ &= \left[\left(I - D_{S_l}\right) - \left(L + E_{S_l}\right)\right]^{-1} \left(U - S_l + S_l U + F_{S_l}\right).\end{aligned}$$

For $A = (a_{ij})$ and $B = (b_{ij}) \in R^{n\times n}$, we write $A \geq B$ whenever $a_{ij} \geq b_{ij}$ holds for all $i, j = 1, 2, \cdots, n$. A is nonnegative if $A \geq 0, (a_{ij} \geq 0; i, j = 1, \cdots, n)$, and $A \geq B$ if and only if $A - B \geq 0$.

Definition 2.1

(Young, [15]). A real n×n matrix $A = (a_{ij})$ with $a_{ij} \leq 0$ for all $i \neq j$ is called a Zmatrix.

Definition 2.2

(Varga, [16]). A matrix A is irreducible if the directed graph associated to A is strongly connected.

Lemma 2.3.

If A is an irreducible diagonally dominant Z-matrix with unit diagonal, and if the assumption (6) holds, then the preconditioned matrix A_S is a diagonally dominant Z-matrix.

Proof. The elements $a_{ij}^{S_l}$ of A_{s_l} are given by

$$a_{ij}^{S_l} = \begin{cases} a_{ij} - a_{k_j, i+1} a_{i+1, j} - a_{i, l_i} a_{l_i, j} & 1 \leq i < n-1, \\ a_{ij} - a_{i, i+1} a_{i+1, j} & i = n-1, \\ a_{ij} & i = n. \end{cases} \tag{7}$$

Since A is a diagonally dominant Z-matrix, so we have

$$\begin{aligned} &0 \le a_{k_i,i+1} a_{i+1,j} \le 1 && \text{for } j \ne i+1, \\ &0 \le a_{i,l_i} a_{l_i,j} \le 1 && \text{for } j \ne l_i, \\ &-1 \le a_{k_i,i+1} a_{i+1,i+1} \le 0. \end{aligned} \tag{8}$$

Therefore, the following inequalities hold:

$$p_i = a_{k_i,i+1} a_{i+1,i} \ge 0,$$

$$q_i = a_{i,l_i} a_{l_i,i} \ge 0,$$

$$t_i = a_{k_i,i+1} \sum_{j=i+1}^{n} a_{i+1,j} \le 0,$$

$$s_i = a_{i,l_i} \sum_{j=1}^{i-1} a_{l_i,j} \ge 0,$$

$$t_i = a_{k_i,i+1} \sum_{j=i+1}^{n} a_{i+1,j} \le 0,$$

$$u_i = a_{i,l_i} \sum_{j=i+1}^{n} a_{l_i,j} \le 0.$$

We denote that $p_n = q_n = r_n = s_n = t_n = u_n = 0$. Then the following inequality holds:

$$p_i + q_i + r_i + s_i + t_i + u_i$$
$$= a_{k_i,i+1} \sum_{j=1}^{n} a_{i+1,j} + a_{i,l_i} \sum_{j=1}^{n} a_{l_i,j} \leq 0 \quad 1 \leq i < n.$$

Furthermore, if $a_{i,i+1} \neq 0$, and $\sum_{j=1}^{n} a_{i,i+1} < 0$, for some i<n, then we have

$$p_i + q_i + r_i + s_i + t_i + u_i < 0 \quad \text{for some } i < n. \tag{9}$$

Let $d(S_l)_i$, $l(S_l)_i$ and $u(S_l)_i$ be the sums of the elements in row i of D_{s_l}, L_{s_l}, and U_{s_l}, respectively. The following equations hold:

$$\begin{aligned} d(S_l) &= a_{ij}^{S_l} = 1 - p_i - q_i, & 1 \leq i \leq n, \\ l(S_l) &= -\sum_{j=1}^{i-1} a_{ij}^{S_l} = l_i + r_i + s_i & 1 \leq i \leq n, \\ u(S_l) &= -\sum_{j=i+1}^{n} a_{ij}^{S_l} = u_i + t_i + u_i & 1 \leq i \leq n, \end{aligned} \tag{10}$$

where l_i and u_i are the sums of the elements in row i of L and U for $A = I - L - U$, respectively. Since A is a diagonally dominant Z-matrix, by (8) and by the condition (6) the following relations hold:

$$1 - a_{k_i,i+1} a_{i+1,j} - a_{i,l_i} a_{l_i,j} > 0, \quad \text{for } j = i$$

$$a_{ij} - a_{k_i,i+1} \sum_{j=i+1}^{n} a_{i+1,j} - a_{i,l_i} \sum_{j=i+1}^{n} a_{l_i,j} \leq 0, \quad \text{for } i > j$$

$$a_{ij} - a_{k_i,i+1} - a_{i,l_i} a_{l_i,i+1} - a_{k_i,i+1} \sum_{j=i+2}^{n} a_{i+1,j} - a_{i,l_i} \sum_{j=i+2}^{n} a_{l_i,j} \leq 0$$

for $i < j$.

Therefore, $l(S_l) \geq 0$, $u(S_l) \geq 0$, and A_{s_l} is a Zmatrix. Moreover, by (9) and by the assumption, we can easily obtain

$$d(S_l) - l(S_l) - u(S_l)$$
$$= (d_i - l_i - u_i) - (p_i + q_i + r_i + s_i + t_i + u_i) > 0 \quad (11)$$

for all i.

Therefore, A_{s_t} satisfies the condition of diagonal dominance.

Lemma 2.4

[10, Lemma 2]. An upper bound on the spectral radius ρ(T) for the Gauss-Seidel iteration matrix T is given by

$$\rho(T) \leq \max_i \frac{\tilde{u}_i}{1 - \tilde{l}_i},$$

where $\tilde{l}_i$ and $\tilde{u}_i$ are the sums of the moduli of the elements in row i of the triangular matrices L and U, respectively.

Theorem 2.5.

Let A be a nonsingular diagonally dominant Z-matrix with unit diagonal elements and let the condition (6) holds, then $\rho(T_{s_t}) < 1$.

Proof. From (11) and $u(S_l) \geq 0$ we have

$$d(S_l) - l(S_l) > u(S_l) \geq 0 \; for\, all\, i.$$

This implies that

$$\frac{u(S_l)}{d(S_l) - l(S_l)} < 1. \quad (12)$$

Hence, by Lemma (2.4) we have $\rho(T_{s_l}) < 1$.

Definition 2.6.

Let A be an n×n real matrix. Then, $A = M - N$ is referred to as:

1. a regular splitting, if M is nonsingular, $M^{-1} \geq 0$ and $N \geq 0$
2. a weak regular splitting, if M is nonsingular, $M^{-1} \geq 0$ and $M^{-1}N \geq 0$.
3. a convergent splitting, if $\rho(M^{-1}N) < 1$.

Lemma 2.7 (Varga, [10]).

Let $A \in R^{n \times n}$ be a nonnegative and irreducible $n \times n$ matrix. Then

1. A has a positive real eigenvalue equal to its spectral radius $\rho(A)$;
2. for $\rho(A)$, there corresponds an eigenvector x > 0;
3. $\rho(A)$ is a simple eigenvalue of A;
4. $\rho(A)$ increases whenever any entry of A increases.

Corollary 2.8 [16, Corollary 3.20].

If $A = (a_{ij})$ is a real, irreducibly diagonally dominant $n \times n$ matrix with $a_{ij} < 0$ for all $i \neq j$, and $a_{ij} > 0$ for all $1 \leq i \leq n$, then $A^{-1} > 0$.

Theorem 2.9 [16, Theorem 3.29]. Let A=MN be a regular splitting of the matrix A. Then, A is nonsingular with $A^{-1} > 0$ if and only if $\rho(M^{-1}N) < 1$, where

$$\rho\left(M^{-1}N\right) = \frac{\rho\left(A^{-1}N\right)}{1 + \rho\left(M^{-1}N\right)} < 1$$

Theorem 2.10 (Gunawardena et al. [2, Theorem 2.2]). Let A be a nonnegative matrix. Then 1) If $ax \leq Ax$ for some nonnegative vector x, $x \neq 0$, then $\alpha \leq \rho(A)$.

2) If $Ax \leq \beta x$ for some positive vector x, then $\rho(A) \leq \beta$ Moreover, if A is irreducible and if $0 \neq \alpha x \leq Ax \leq \beta x$ for some nonnegative vector x, then $\alpha \leq \rho(A) \leq \beta$ and x is a positive vector.

Let B be a real Banach space, B its dual and L(B) the space of all bounded linear operator mapping B into itself. We assume that B is generated by a normal cone K [17]. As is defined in [17], the operator $A \in L(B)$ has the property "d" if its dual A′ possesses a Frobenius eigenvector in the dual cone K′ which is defined by

$$K' = \{x' \in B' : \langle x, x' \rangle = x'(x) \geq 0 \quad \text{for all } x \in K\}.$$

As is remarked in [1,17], when $B = R^n$ and $K = R^n_+$, all $n \times n$ real matrices have the property "d". Therefore the case are discussing fulfills the property "d". For the space of all $n \times n$ matrices, the theorem of Marek and Szyld can be stated as follows:

Theorem 2.11 (Marek and Szyld [17, Theorem 3.15]). Let $A_1 = M_1 - N_1$ and $A_2 = M_2 - N_2$ be weak regular splitting with $T_1 = M_1^{-1}N_1, T_2 = M_2^{-1}N_2$. Let $x \geq 0, y \geq 0$ be such that $T_1 x = \rho(T_1)x$ and $T_2 y = \rho(T_2)y$. If $M_1^{-1} \geq M_2^{-1}$ and if either $(A_1 - A_2)x \geq 0, A_1 x \geq 0$, or $(A_1 - A_2)y \geq 0, A_1 y \geq 0$, with $y > 0$, then $\rho(T_1) \leq \rho(T_2)$

Moreover, if $M_1^{-1} > M_2^{-1}$ and $N_1 \neq N_2$ then $\rho(T_1) < \rho(T_2)$

Now in the following lemma we prove that $A_{s_l} = (I + S_n + S_m)A = M_{s_l} - N_{s_l}$ is Gauss-Seidel convergent regular splitting.

Theorem 2.12. Let A be an irreducibly diagonally dominant Z-matrix with unit diagonal, and let the condition (6) holds, then $A_{s_l} = M_{s_l} - N_{s_l}$ is Gauss-Seidel convergent regular splitting. Moreover

$$\rho(T_{S_l}) \leq \rho(T_{s+m}) < 1.$$

Proof. If A is an irreducibly diagonally dominant Zmatrix, then by Lemma (2.3), A_{s_l} is a diagonally dominant Z -matrix. So we have $A_{s_l}^{-1} \geq 0$. By hypothesis we have $(I - D_{s_l})^{-1} \geq I$. Thus the strictly lower triangular matrix $(L + E_{s_l})$ has nonnegative elements. By considering Neumanns series, the following inequality holds:

$$M_{S_l}^{-1} = \Big[I + \left(I - D_{S_l}\right)^{-1}\left(L + E_{S_l}\right) + \left\{\left(I - D_{S_l}\right)^{-1}\left(L + E_{S_l}\right)\right\}^2 + \cdots + \left\{\left(I - D_{S_l}\right)^{-1}\left(L + E_{S_l}\right)\right\}^{n-1}\Big]\left(I - D_{S_l}\right)^{-1} \geq 0.$$

Direct calculation shows that $N_{s_l} \geq 0$ holds. Thus, by definition (2.6) $A_{s_l} = M_{s_l} - N_{s_l}$ is the Gauss-Seidel convergent regular splitting. Also in [3] we have $A_{s+m} = M_{s+m} - N_{s+m}$ and

$$M_{s+m}^{-1} = [I + \left(I - D_{S+M}\right)^{-1}\left(L + E_{S+M}\right) + \left\{\left(I - D_{S+M}\right)^{-1}\left(L + E_{S+M}\right)\right\}^2 + \cdots, + \left\{\left(I - D_{S+M}\right)^{-1}\left(L + E_{S+M}\right)\right\}^{n-1}\Big]\left(I - D_{S+M}\right)^{-1} \geq 0.$$

Direct comparison of the two matrix elements $(I - D_{s+m})^{-1}$ and $(I - D_{s_l})^{-1}$ also $(L + E_{S+M})$ and $(L + E_{s_l})$ we obtain

$$\left(I - D_{S+M}\right)^{-1} \leq \left(I - D_{S_l}\right)^{-1},$$
$$\left(L + E_{S+M}\right) \leq \left(L + E_{S_l}\right).$$

Thus

$$0 \leq M_{s+m}^{-1} \leq M_{s_l}^{-1}$$

Furthermore, since $S_n \geq S$, we have $A_{s_l}x - A_{s+m}x = (S_n - S)Ax \geq 0$. From Lemma (2.7), x is an eigenvector of T_{s_l}, and x is also a Perron vector of T_{s_l}. Therefore, from Theorem (2.11), $\rho(T_{s_l}) \leq \rho(T_{s+m})$ holds.

Denote

$$
\begin{aligned}
A_{smr} &= \left(I + S + S_m + R\right)A, \\
A_{S_l r} &= \left(I + S_l + R\right)A, \\
A_G &= \left(I + S + S_m + \gamma G\right)A, \\
A_{GS_l} &= \left(I + S_l + \gamma G\right)A,
\end{aligned}
$$

and also let T_{smr}, $T_{s_l r}$, T_G and $T_G S_l$ be the iterative matrix associated to $A_{smr}, A_{s_l r}, A_G$ and A_{GS_l} respectively. Then we can prove $\rho(T_{s_l r}) \le \rho(T_{smr})$ and $\rho(T_G S_l) \le \rho(T_G)$, similarly. In summary, we have the following inequalities:

$$
\begin{aligned}
\rho(T) &\ge \rho(T_s) \ge \rho(T_{s+m}) \ge \rho\left(T_{S_l}\right) \ge \rho\left(T_{smr}\right) \\
&\ge \rho\left(T_{S_l} r\right) \ge \rho(T_G) \ge \rho\left(T_{GS_l}\right).
\end{aligned}
$$

Remark 2.13. W. Li, in [18] used the M-matrix instead of irreducible diagonally dominant Z-matrix, therefore we can say that the Lemma 2.3 and the Theorems 2.5 and 2.12 are hold for M-matrices.

NUMERICAL RESULTS

In this section, we test a simple example to compare and contrast the characteristics of the different preconditioners. Consider the matrix

$$
A = \begin{pmatrix} 1 & -0.2 & -0.6 & -0.2 \\ -0.2 & 1 & -0.3 & -0.1 \\ -0.1 & -0.2 & 1 & -0.3 \\ -0.2 & -0.3 & -0.2 & 1 \end{pmatrix}
$$

Applying the Gauss-Seidel method, we have $\rho(T) = .0.5099$. By using preconditioner $P_s = (I+S)A$, we find that A_s and T_s have the following forms:

$$A_S = \begin{pmatrix} 0.96 & 0 & -0.66 & -0.22 \\ -0.23 & 0.94 & 0 & -0.19 \\ -0.16 & -0.29 & 0.94 & 0 \\ -0.2 & -0.3 & -0.2 & 1 \end{pmatrix}$$

$$T_S = \begin{pmatrix} 0 & 0 & 0.6875 & 0.2292 \\ 0 & 0 & 0.1682 & 0.2582 \\ 0 & 0 & 0.1689 & 0.1187 \\ 0 & 0 & 0.2217 & 0.1470 \end{pmatrix}$$

And $\rho(T_s) = 0.3205$.

Using the preconditioner P_{s+m} we obtain

$$A_{s+m} = \begin{pmatrix} 0.90 & -0.12 & -0.06 & -0.4 \\ -0.25 & 0.91 & -0.2 & -0.09 \\ -0.16 & -0.29 & 0.94 & 0 \\ -0.2 & -0.3 & -0.2 & 1 \end{pmatrix}$$

$$T_{s+m} = \begin{pmatrix} 0 & 0.133 & 0.067 & 0.444 \\ 0 & 0.037 & 0.040 & 0.221 \\ 0 & 0.034 & 0.023 & 0.144 \\ 0 & 0.044 & 0.030 & 0.184 \end{pmatrix}$$

and $\rho(T_{s+m}) = 0.2581$.

For $P_{s_l} = I + S_l$, we have

$$A_{S_l} = \begin{pmatrix} 0.88 & -0.02 & -0.09 & -0.41 \\ -0.25 & 0.91 & -0.2 & -0.09 \\ -0.16 & -0.29 & 0.94 & 0 \\ -0.2 & -0.3 & -0.2 & 1 \end{pmatrix}$$

$$T_{S_l} = \begin{pmatrix} 0 & 0.023 & 0.102 & 0.465 \\ 0 & 0.006 & 0.050 & 0.227 \\ 0 & 0.006 & 0.033 & 0.149 \\ 0 & 0.007 & 0.042 & 0.1911 \end{pmatrix}$$

And $\rho(T_{s_l}) = 0.2328$.

For $P_{smr} = I + S + S_m + R$, we have

$$A_{smr} = \begin{pmatrix} 0.90 & -0.12 & -0.06 & -0.4 \\ -0.25 & 0.91 & -0.2 & -0.09 \\ -0.16 & -0.29 & 0.94 & 0 \\ -0.08 & -0.08 & -0.21 & 0.87 \end{pmatrix}$$

$$T_{smr} = \begin{pmatrix} 0 & 0.133 & 0.067 & 0.444 \\ 0 & 0.037 & 0.040 & 0.221 \\ 0 & 0.034 & 0.023 & 0.144 \\ 0 & 0.024 & 0.015 & 0.096 \end{pmatrix}$$

and $\rho(T_{smr}) = 0.1694$.

For $P_{s_l r} = I + S_l + R$, we have

$$A_{S_l r} = \begin{pmatrix} 0.88 & -0.02 & -0.09 & -0.41 \\ -0.25 & 0.91 & -0.02 & -0.09 \\ -0.16 & -0.29 & 0.94 & 0 \\ -0.08 & -0.08 & -0.21 & 0.87 \end{pmatrix}$$

$$T_{S_l r} = \begin{pmatrix} 0 & 0.023 & 0.102 & 0.465 \\ 0 & 0.006 & 0.050 & 0.227 \\ 0 & 0.006 & 0.033 & 0.149 \\ 0 & 0.004 & 0.022 & 0.1 \end{pmatrix}$$

and $\rho(T_{s_l r}) = 0.1415$

From the above results, we have G_{nj}=(0.28,0.38,0.41,0). Then $A_G(\gamma = 1)$ and T_G have the forms:

$$A_G = \begin{pmatrix} 0.90 & -0.12 & -0.06 & -0.4 \\ -0.25 & 0.91 & -0.02 & -0.09 \\ -0.16 & -0.29 & 0.94 & 0 \\ -0.04 & -0.06 & -0.07 & 0.78 \end{pmatrix}$$

$$T_G = \begin{pmatrix} 0 & 0.133 & 0.067 & 0.444 \\ 0 & 0.037 & 0.040 & 0.221 \\ 0 & 0.034 & 0.024 & 0.144 \\ 0 & 0.012 & 0.008 & 0.050 \end{pmatrix}$$

and $\rho(T_G) = 0.1218$.

For $P_{GSl}(\gamma = 1) = I + S_l + G$, we have

$$A_{GS_l} = \begin{pmatrix} 0.88 & -0.02 & -0.09 & -0.41 \\ -0.25 & 0.91 & -0.02 & -0.09 \\ -0.16 & -0.29 & 0.94 & 0 \\ -0.04 & -0.06 & -0.07 & 0.78 \end{pmatrix}$$

$$T_{GS_l} = \begin{pmatrix} 0 & 0.023 & 0.102 & 0.465 \\ 0 & 0.006 & 0.050 & 0.227 \\ 0 & 0.006 & 0.033 & 0.149 \\ 0 & 0.002 & 0.011 & 0.052 \end{pmatrix}$$

and $\rho(T_{GS_l}) = 0.0940$. Since the preconditioned matrices differ only in the values of their last rows, the related matrices also differ only in these values, as is shown in the above results. Thus the elements of new A_G and A_{GS_l} are similar to elements of A_G and A_{GS_l}, respectively than the elements of last rows. Therefore, we hereafter show only the last row.

By putting $\gamma_1 = 1.22699$, the matrices $A_{GS_{l\gamma l}}$ $T_{G_{\gamma l}}$ $A_{G_{\gamma l}}$ and $T_{GS_{l\gamma l}}$ have the following forms:

$$A_{G_{\gamma_1}} = (0, -0.003, -0.042, 0.733),$$
$$T_{G_{\gamma_1}} = (0, 0.0021, 0.0015, 0.0093),$$

and $\rho(T_{G_{\gamma 1}}) = 0.0779$,

$$A_{GS_l\gamma_1} = (0, -0.003, -0.042, 0.733),$$
$$T_{GS_l\gamma_1} = (0, 0.00036, 0.0021, 0.0096),$$

and $\rho(T_{GS_{l\gamma 1}}) = 0.0509$.

For $\gamma_2 = 1.23966$ we have:

$$A_{G_{\gamma_2}} = (0.0021, 0, -0.0413, 0.7310),$$
$$T_{G_{\gamma_2}} = (0, 0.0015, 0.0011, 0.0069),$$

and $\rho(T_{G_{\gamma 2}}) = 0.0751$ and

$$A_{GS_l\gamma_2} = (0.0021, 0, -0.0413, 0.7310),$$
$$T_{GS_l\gamma_2} = (0, 0.00026, 0.0016, 0.0071),$$

and $\rho(T_{GS_{l\gamma 2}}) = 0.0483$.

For $\gamma_3(opt) = 1.5625$ we have:

$$A_{G_{\gamma_3}(opt)} = (0.0547, 0.0781, 0, 0.6609),$$
$$T_{G_{\gamma_3}(opt)} = (0, -0.0154, -0.0102, -0.0629),$$

and $\rho\left(T_{G_{\gamma 3(opt)}}\right) = 0.0227$ and

$$A_{GS_l\gamma_3(opt)} = (0.0547, 0.0781, 0, 0.6609),$$
$$T_{GS_l\gamma_3(opt)} = (0, -0.0026, -0.0144, -0.0654),$$

and $\rho\left(T_{GS_{l\gamma 3(opt)}}\right) = 0.0216$.

From the numerical results, we see that this method with the preconditioner $P_{GS_l}(\gamma_{opt})$ produces a spectral radius smaller than the recent preconditioners that above was introduced.

REFERENCES

1. T. Z. Huang, G. H. Cheng and X. Y. Cheng, "Modified SOR-Type Ierative Method for Z-Matrices," Applied Mathematics and Computation, Vol. 175, No. 1, 2006, pp. 258-268.doi:10.1016/j.amc.2005.07.050

2. A. D. Gunawardena, S. K. Jain and L. Snyder, "Modified Iterative Method for Consistent Linear Systems," Linear Algebra and Its Applications, Vol. 154-156, 1991, pp. 123–143.doi:10.1016/0024-3795(91)90376-8
3. M. Morimoto, K. Harada, M. Sakakihara and H. Sawami, "The Gauss-Seidel Iterative Method with the Preconditioning Matrix $(I + S + S_m)$," Japan Journal of Industrial and Applied Mathematics, Vol. 21, No. 1, 2004, pp. 25–34. doi:10.1007/BF03167430
4. D. Noutsos and M. Tzoumas, "On Optimal Improvements of Classical Iterative Schemes for Z-Matrices," Journal of Computational and Applied Mathematics, Vol. 188, No. 1, 2006, pp. 89-106. doi:10.1016/j.cam.2005.03.057
5. H. Niki, T. Kohno and M. Morimoto, "The Preconditioned Gauss-Seidel Method Faster than the SOR Method," Journal of Computational and Applied Mathematics, Vol. 218, No. 1, 2008, pp. 59-71. doi:10.1016/j.cam.2007.07.002
6. H. Niki, T. Kohno and K. Abe, "An Extended GS Method for Dense Linear System," Journal of Computational and Applied Mathematics, Vol. 231, No. 1, 2009, pp. 177-186.doi:10.1016/j.cam.2009.02.005
7. M. Morimoto, H. Kotakemori, T. Kohno and H. Niki, "The Gauss-Seidel Method with Preconditioner (I + R)," Transactions of the Japan Society for Industrial and Applied Mathematics, Vol. 13, 2003, pp. 439-445.
8. H. Kotakemori, K. Harada, M. Morimoto and H. Niki, "A Comparison Theorem for the Iterative Method with the Preconditioner $(I + S_{max})$," Journal of Computational and Applied Mathematics, Vol. 145, No. 2, 2002, pp. 373-378. doi:10.1016/S0377-0427(01)00588-X
9. A. Hadjidimos, D. Noutsos amd M. Tzoumas, "More on Modifications and Improvements of Classical Iterative Schemes for Z-Matrices," Linear Algebra and Its Applications, Vol. 364, 2003, pp. 253-279. doi:10.1016/S0024-3795(02)00570-0
10. T. Kohno, H. Kotakemori, H. Niki and M. Usui, "Improving the Gauss-Seidel Method for Z-Matrices," Linear Algebra and Its Applications, Vol. 267, No. 1-3, 1997, pp. 113-123.
11. W. Li and W. Sun, "Modified Gauss-Seidel Type Methods and Jacobi Type Methods for Z-Matrices," Linear Algebra and Its Applications, Vol. 317, 2000, pp. 227-240.doi:10.1016/S0024-3795(00)00140-3
12. J. P. Milaszewicz, "On Modified Jacobi Linear Operators," Linear Algebra and Its Applications, Vol. 51, 1983, pp. 127-136. doi:10.1016/0024-3795(83)90153-2
13. J. P. Milaszewicz, "Improving Jacobi and Gauss-Seidel Iterations," Linear Algebra and Its Applications, Vol. 93, 1987, pp. 161-170. doi:10.1016/S0024-3795(87)90321-1
14. H. Niki, K. Harada, M. Morimoto and M. Sakakihara, "The Survey of Preconditioners Used for Accelerating the Rate of Convergence in the Gauss-Seidel Method," Journal of Computational and Applied Mathematics, Vol. 164–165, 2004, pp. 587-600.doi:10.1016/j.cam.2003.11.012

15. D. M. Young, "Iterative Solution of Large Linear Systems," Academic Press, New York, 1971.
16. R. S. Varga, "Matrix Iterative Analysis," Prentice-Hall, Englewood Cliffs, 1962.
17. I. Marek and D. Szyld, "Comparison Theorems for Weak Splittings of Bound Operators," Numerische Mathematik, Vol. 58, No. 1, 1990, pp. 387-397. doi:10.1007/BF01385632
18. W. Li, "Comparison Results for Solving Preconditioned Linear Systems," Journal of Computational and Applied Mathematics, Vol. 176, No. 2, 2005, pp. 319-329.doi:10.1016/j.cam.2004.07.022

CITATION

1. A. Nazari and S. Borujeni, "A Modified Precondition in the Gauss-Seidel Method," Advances in Linear Algebra & Matrix Theory, Vol. 2 No. 3, 2012, pp. 31-37. doi: 10.4236/alamt.2012.23005.

Solving the Class Equation $x^d = \beta$ in an Alternating Group for Each $\beta \in C^{a} \cap H_n^c$ and n>1

Shuker Mahmood*[1, 2] *and
***Andrew Rajah*[2]**

[1]Department of Mathematics, College of Science, University of Basra, Basra, Iraq
[2]School of Mathematical Sciences, Universiti Sains Malaysia, Penang, Malaysia

7

ABSTRACT

The main purpose of this paper is to solve the class equation $x^d = \beta$ in an alternating group, (i.e. find the solutions set $X = \{x \in A_n \mid x^d \in A(\beta)\}$) and find the number of these solutions $|X|$ where β ranges over the conjugacy class A (β) in A_n and d is a positive integer. In this paper we solve the class equation $x^d=\beta$ in A_n where $\beta \in H_n^c \cap C^{\alpha}$, for all n>1. H_n^c is the complement set of H_n where $H_n = \{C^{\alpha}$ of $S_n \mid n > 1$, with all parts α_k of α are different and odd}. C^{α} is conjugacy class of S_n and form class C^{α} depends on the cycle type α of its elements If $\lambda \in C^{\alpha}$ and $\lambda \in H_n \cap C^{\alpha}$, then C^{α} splits into the two classes $C^{\alpha\pm}$ of A_n.

INTRODUCTION

The Frobenius equation $x^d = \beta$ in finite groups was introduced by G. Frobenius and then was studied by many others such as ([1-4]). Where they dealt with some types of finite groups like finite cyclic groups, finite pgroups, Wreath products of finite groups, etc. Choose any $\beta \in S_n$

and write it as $\lambda_1\lambda_2...\lambda_c(\beta)$. With γ_i disjoint cycles of length α_i and c(β) is the number of disjoint cycle factors including the 1-cycle of β. Since disjoint cycles commute, we can assume that $\alpha_1 \geq \alpha_2 \geq ...\alpha_{c(\beta)}$. Therefore $\alpha = (\alpha_1, \alpha_2, ..., \alpha_{c(\beta)})$ is a partition of n and it is call cycle type of β. Let $C^a \subset S_n$ be the set of all elements with cycle typeα, then we can determine the conjugate class of $\beta \in S_n$ by using cycle type of β, since each pair of λ and β in S_n are conjugate if they have the same cycle type (see [5]). Therefore, the number of conjugacy classes of S_n is the number of partitions of n. However, this is not necessarily true in an alternating group. Let β= (124) and λ=(142) are two permutations in S_4 we have they are belong to the same conjugate class C^α=[1,3] in S_4 (i.e. $C^\alpha(\beta)=C^\alpha(\lambda)$) since

$$\alpha(\beta) = (\alpha_1(\beta), \alpha_2(\beta)) = (1,3)$$
$$= (\alpha_1(\lambda), \alpha_2(\lambda)) = \alpha(\lambda)$$

that means they have the same cycle type but in fact λ and β are not conjugate inA_4, also let β=(123)(456)(789) and λ=(537)(169)(248) in S_9 we have they are belong to the same conjugate class $C^\alpha=[3^3]$ in S_4 since α(β)=(3,3,3)=α(λ) but here they are conjugate inA_9. So from the first and second examples we consider it is not necessarily if two permutations have the same cycle type are conjugate in A_n therefore in this work we discuss in detail the conjugacy classes in an alternating group and we denote to conjugacy class of β in A_n by A(β). Also we introduce some theorems to solve the class equation x^d= β in A_n where $\beta \in H_n^c \cap C^\alpha$, for all n>1.

Definition [6]

A partition α is a sequence of nonnegative integers (α_1, α_2,...) with $\alpha_1 \geq \alpha_2 \geq ...$ and $\sum_{i=1}^{\infty} \alpha_i < \infty$. The length l(α)and the size |α| of α are defined as

$$l(\alpha) = \max\{i \in N; \alpha_i \neq 0\}$$

and $|\alpha| = \sum_{i=1}^{\infty} \alpha_i$. We set α n={α partition; |α|=n} for $n \in N$. An element of α n is called a partition of n.

Remark [6]

We only write the non zero components of a partition. Choose any $\beta \in S_n$ and write it as $\lambda_1\lambda_2 \ldots \lambda_c(\beta)$. With γ_i disjoint cycles of length α_i and c(β) is the number of disjoint cycle factors including the 1-cycle of β. Since disjoint cycles commute, we can assume that $\alpha_1 \geq \alpha_2 \geq \ldots \geq \alpha_{c(\beta)}$. Therefore $\alpha = (\alpha_1, \alpha_2, \ldots, \alpha_{c(\beta)})$ is a partition of n and each α_i is called part of α.

Definition [6]

We call the partition

$$\alpha = \alpha(\beta) = \left(\alpha_1(\beta), \alpha_2(\beta), \cdots, \alpha_{c(\beta)}(\beta)\right)_i$$

the cycle type of β.

Definition [6]

Let α be a partition of n. We define $C^\alpha \subset S_n$ to be the set of all elements with cycle type α.

Definition [6]

Let $\beta \in S_n$ be given. We define $c_m = c_m^{(n)} = c_m^{(n)}(\beta)$ to be the number of cycles of length m of β.

Remarks

- If $\beta \in C^{\alpha}$, then we write $C^{\alpha} = C^{\alpha}(\beta)$.
- The relationship between partitions and c_m is as follows: if $\beta \in C^{\alpha}$ is given then $c_m^{(n)}(\beta) = |\{i : \alpha_i = m\}|$, (see [6])
- The cardinality of each $C^{\alpha} = C^{\alpha}(\beta)$ can be found as follows: $|C^{\alpha}| = \frac{n!}{z_{\alpha(\beta)}}$ with $z_{\alpha(\beta)} = \prod_{r=1}^{n} r^{c_r}(c_r)$ and $c_r = c_r^{(n)}(\beta) = |\{i : \alpha_i = r\}|$, (see [7]).
- $C^{\alpha}(\beta)$ splits into two A_n-classes of equal order iff n>1, and the non-zero parts of $\alpha(\beta)$ are different and odd, in every other case $C^{\alpha}(\beta)$ does not split, (see [8]).

Lemma [9]

Let p prime number and $[a^r]$ a conjugate class of symmetric group. If p does not divide a, then the solutions of $x^p \in [a^r]$ are:

- $[a^r]$, if $1 \le r < p$
- $[a^r], [(pa), a^{r-p}], [(pa)^2, a^{r-2p}], \cdots, [(pa)^m, a^{r-mp}]$ if mp$\le$r<(m+1)p

Lemma [9]

Let p and q be different prime numbers and $[a^r]$ a conjugate class of symmetric group. If p|a and q does not divide a, then the solutions of $x^{pq} \in [a^r]$ are:

- $[(pa)^i, (pqa)^j]$, where i and j are solutions of the equation $i + qj = \frac{r}{p}$ if p|r.
- No solution if p does not divide r.

Lemma [9]

Let p and q be different prime numbers and $\left[a^r\right]$ a conjugate class in S_n. If p does not divide a and q does not divide a, then the solutions of $x^{pq} \in \left[a^r\right]$ are $\left[a^i,(pa)^j,(qa)^k,(pqa)^l\right]$, where i, j, k and l are non-negative integers and solutions of the equation i + pj + qk + pql = r.

CONJUGACY CLASS A (β) 2075 /> OF AN [10]

Let $\beta \in C^\alpha$, where β is a permutation in an alternating group. We define the A (β) conjugacy class of β in A_n by:

$$A(\beta)=\left\{\gamma \in A_n \middle| \gamma = t\beta t^{-1}; \text{for some } t \in A_n\right\}$$

$$=\begin{cases} C^\alpha, & \text{if } \beta \notin H_n \\ C^{\alpha+} \text{ or } C^{\alpha-}, & \text{if } \beta \in H_n \end{cases}$$

where H_n={ Cα of S_n|n>1, with all parts α_k of α different and odd}.

Remarks

- $\beta \in H_n \Rightarrow \beta \in A_n$ ..
- $\beta \in C^\alpha \cap H_n^c \cap A_n \Rightarrow A(\beta)=C^\alpha$ where H_n^c is complement of H_n.
- $\beta \in C^\alpha \cap H_n \Rightarrow \beta \in A_n$ and C^α split into two classes $C^{\alpha\pm}$ of A_n.
- If $\beta,\lambda \in C^\alpha \cap H_n$, and $\lambda \in C^{\alpha\pm}$, then

$$A(\beta)=\begin{cases} C^{\alpha+} & \text{if } \beta \underset{A_n}{\approx} \lambda \\ C^{\alpha-} & O \cdot W \end{cases}$$

— If $n \in \theta = \{1,2,5,6,10,14\}$, then for each $\beta \in A_n$, β is conjugate to β^{-1} in $A_n \left(\beta \underset{A_n}{\approx} \beta^{-1} \right)$.

Definition

Let F_n= { C^α of S_n| the number of parts α_k of α with the property $\alpha_k \equiv 3$ (mod 4) is odd}. Then, for each $\beta \in H_n \cap C^\alpha \cap F_n$, $^{\alpha\pm}$ of A_n is defined by

$$C^{\alpha+} = \left\{ \lambda \in A_n \middle| \lambda = \gamma\beta\gamma^{-1}; \text{for some } \gamma \in A_n \right\} = A(\beta),$$

$$C^{\alpha-} = \left\{ \lambda \in A_n \middle| \lambda = \gamma\beta^{-1}\gamma^{-1}; \text{for some } \gamma \in A_n \right\} = A\left(\beta^{-1}\right).$$

Definition

Let $\overline{F_n}$ = {C^α of S_n| the number of parts α_k of α with the property $\alpha_k \equiv 3$ (mod 4) is even}. Then, for each $\beta \in H_n \cap C^\alpha \cap \overline{F_n}$, $C^{\alpha\pm}$ of A_n is defined by

$$C^{\alpha+} = \left\{ \lambda \in A_n \middle| \lambda = \gamma\beta\gamma^{-1}; \text{for some } \gamma \in A_n \right\} = A(\beta),$$

$$C^{\alpha-} = \left\{ \lambda \in A_n \middle| \lambda = \gamma\beta^{-1}\gamma^{-1}; \text{for some } \gamma \in A_n \right\} = A\left(\beta^{\#}\right),$$

where $\beta^{\#}$ does not conjugate to β.

RESULTS FOR EVEN PERMUTATIONS IN H_n^c

Theorem

Let A (β) be the conjugacy class of β in A_n. If p is a prime number and does not divide a, $\beta \in \left[a^r\right] \cap H_n^c$, where $\left[a^r\right]$ is a class of S_n, then the solutions of $x^p \in A(\beta)$ are

1. $\left[a^r\right]$ if $(1 \le r < p)$ and (a is odd or (a and r) are even)
2. $\left[a^r\right], \left[(pa), a^{r-p}\right], \left[(pa)^2, a^{r-2p}\right], \cdots, \left[(pa)^m, a^{r-mp}\right]$

   if [((a and p) are odd) or (p is odd and (a and r) are even)] and $\left[mp \le r < (m+1)p\right]$
3. $\left[a^r\right], \left[(pa)^2, a^{r-2p}\right], \left[(pa)^4, a^{r-4p}\right], \cdots,$
4. $\left[(pa)^m, a^{r-mp}\right]$

   if [(a is odd and p is even) or (a, p and r are even)] and $[mp \le r < (m+1)p$ and m is even]
5. $\left[a^r\right], \left[(pa)^2, a^{r-2p}\right], \left[(pa)^4, a^{r-4p}\right], \cdots,$

   $\left[(pa)^{(m-1)}, a^{r-(m-1)p}\right]$

   if [(a is odd and p is even) or (a, p and r are even)] and [$mp \le r < (m+1)p$ and m is odd]
6. $\left[(pa), a^{r-p}\right], \left[(pa)^3, a^{r-3p}\right], \cdots, \left[(pa)^m, a^{r-mp}\right]$ if

   [(a and p) are even and r is odd] and $[mp \le r < (m+1)p$ and m is odd]
7. $\left[(pa), a^{r-p}\right], \left[(pa)^3, a^{r-3p}\right], \cdots, \left[(pa)^{(m-1)}, a^{r-(m-1)p}\right]$

   if [(a and p) are even and r is odd)] and $[mp \le r < (m+1)p$ and m is even], or
8. Does not exist, if [(a is even and (p and r) are odd].

Proof

Given that $\beta \in [a^r] \cap H_n^c \cap A_n$, A ($\beta$)= $[a^r]$, then by (1.7), the solutions of $x^p \in A(\beta)$ in S_n are

a. $[a^r]$, if $1 \le r < p$, or $[a^r]$

b. $[(pa), a^{r-p}], [(pa)^2, a^{r-2p}], [(pa)^m, a^{r-mp}]$ if $mp \le r < (m+1)p$.

2. Assume $(1 \le r < p)$ and (a is odd or (a and r) are even), then from a), $[a^r]$ is the solution set of $x^p \in A(\beta)$ in S_n. Let $\lambda \in [a^r]$. If a is odd and $\lambda = \lambda_1\lambda_2 \ldots \lambda_r$ where $[\langle \gamma_i \rangle] = a$ (odd) for each $(1 \le i \le r)$, then γ_i is a product of an even number similar to T_i of transpositions for all $(1 \le i \le r)$. For any r (odd or even), λ is a product of $(T_1+T_2+\ldots+T_r)$ = (even) number of transpositions $\Rightarrow \lambda \in A_n$. If a and r are even and $\lambda = \gamma_1\gamma_2 \ldots \gamma_r$, where $[\langle \gamma_i \rangle] = a$ (even) for each $(1 \le i \le r)$, then γ_i is a product of an odd number similar to T_i of transpositions for all $(1 \le i \le r) \Rightarrow \lambda$ is a product of $(T_1+T_2+\ldots+T_r)$= (even) number of transpositions $\Rightarrow \lambda \in A_n$, then the solution set of $x^p \in A(\beta)$ in A_n is $[a^r]$.

2. Assume [(a and p) are odd) or (p is odd and (a and r) are even)] and $mp \le r < (m+1)p$ then from b), $[a^r], [(pa), a^{r-p}], [(pa)^2, a^{r-2p}], \ldots, [(pa)^m, a^{r-mp}]$ are solutions of $x^p \in A(\beta)$ in S_n. Let $\lambda \in [a^r] \Rightarrow \lambda \in A_n$, considering that (a is odd or (a and r) are even) and for each $\lambda \notin [(pa)^k, a^{r-kp}]$ $(1 \le k \le m)$. If a and p are odd, then $\lambda = \mu\gamma$, where $\gamma = \gamma_1\gamma_2 \ldots \gamma_{r-kp}$ and $\forall (1 \le i \le r-kp) \Rightarrow \gamma_i \, |\langle \gamma_i \rangle| = a(\text{odd})$ is a product of an even number of transpositions for all, $(1 \le i \le r-kp) \Rightarrow \gamma$ is a product of an even number of transpositions, and $\mu = \mu_1\mu_2 \ldots \mu_k$, where $|\langle \mu_i \rangle| = ap$ (odd), $\forall (1 \le j \le k) \Rightarrow \mu_j$ is a product of an even number of transpositions for all and $(1 \le j \le k) \Rightarrow \mu$ is a product of an even number of transposi-

tions$\Rightarrow \lambda \in A_n$. If (p is odd and (a and r) are even), then γ_i is a product of an odd number similar to L_i of transpositions, $\forall(1 \le i \le r-kp)$. Moreover, $|\langle \mu_i \rangle| = ap$ (even), and $\forall(1 \le j \le kp) \Rightarrow \mu_j$ is a product of an odd number similar to T_j of transpositions for all $(1 \le j \le k)$. If k is odd, then λ is a product of $(L_1+L_2+\ldots+L_{r-kp}) + (T_1+T_2+\ldots+T_k)$ = (odd) + (odd) = (even) number of transpositions $\Rightarrow \lambda \in A_n$. If k is even, then λ is a product of $(L_1+L_2+\ldots+L_{r-kp}) + (T1+T_2+\ldots+T_k)$ = (even) + (even) = (even) number of transpositions $\Rightarrow \lambda \in A_n$, then the solutions of $x^p \in A(\beta)$ in A_n are

$$\left[a^r\right],\left[(pa),a^{r-p}\right],\left[(pa)^2,a^{r-2p}\right],\cdots,\left[(pa)^m,a^{r-mp}\right].$$

3. and 4) Assume [(a is odd and p is even) or (a, p, and r are even)] and $(mp \le r < (m+1)p)$. Then, from b), $\left[a^r\right]$, $\left[(pa),a^{r-p}\right]$, and $\left[(pa)^2,a^{r-2p}\right],\ldots,\left[(pa)^m,a^{r-mp}\right]$ are solutions of $x^p \in A(\beta)$ in S_n. Let $\lambda \in \left[a^r\right] \Rightarrow \lambda \in A_n$, [a is odd or (a and r) are even]. For each $\lambda \in \left[(pa)^k,a^{r-kp}\right]$, $(1 \le k \le m)$ if (a is odd and p is even) $\Rightarrow \lambda = \mu\gamma$, where $\gamma = \gamma_1\gamma_2\ldots\gamma_{r-kp}$ and $\left[\langle \gamma_i \rangle\right] = a$ (odd), $\forall(1 \le i \le r-kp) \Rightarrow \gamma_i$ is a product of an even number similar to L_i of transpositions, $(1 \le i \le r-kp)$ and $\mu=\mu_1\mu_2\ldots\mu_k$, where $|\langle \mu_i \rangle| = ap$ (even), $\forall(1 \le j \le k) \Rightarrow \mu_j$ is a product of an odd number similar to T_j of transpositions for all $(1 \le j \le k)$. If k is odd, then λ is a product of $(L_1+L_2+\ldots+L_{r-kp}) + (T_1+T_2+\ldots+T_k)$ = (even) + (odd) = (odd) number of transpositions$\Rightarrow \lambda \notin A_n$. If k is even, then λ is a product of $(L_1+L_2+\ldots+L_{r-kp})$+ $(T_1+T_2+\ldots+T_k)$ = (even) + (even) = (even) number of transpositions$\Rightarrow \lambda \in A_n$. If (a, p, and r are even), then $\lambda = \mu\gamma$, where $\gamma = \gamma_1\gamma_2\ldots\gamma_{r-kp}$ and $|\langle \gamma_i \rangle = a|$

(even), $\forall(1\le i\le r-kp)\Rightarrow\gamma_i$ is a product of an odd number similar to L_i of transpositions, $(1\le i\le r-kp)$ and $\mu=\mu_1\mu_2\ldots\mu_k$, where $|\langle\mu_i\rangle|=ap$ (even), $\forall(1\le j\le k)\Rightarrow\mu_j$ is a product of an odd number similar to T_j of transpositions for all $(1\le j\le k)$. If k is odd, then λ is a product of $(L_1+L_2+\ldots+L_{r-kp})$ + $(T_1+T_2+\ldots+T_k)$ = (even) + (odd) = (odd) number of transpositions $\Rightarrow\lambda\in A_n$. If k is even, then λ is a product of $(L_1+L_2+\ldots+L_{r-kp})$ + $(T_1+T_2+\ldots+T_k)$ = (even) + (even) = (even) number of transpositions $\Rightarrow\lambda\in A_n$, then the solutions of $x^p\in A(\beta)$ in A_n are $[a^r],[(pa)^2,a^{r-2p}],[(pa)^4,a^{r-4p}],\ldots,[(pa)^m,a^{r-mp}]$ (if m is even) and

$$\left[a^r\right],\left[(pa)^2,a^{r-2p}\right],\left[(pa)^4,a^{r-4p}\right],\cdots,$$
$$\left[(pa)^{(m-1)},a^{r-(m-1)p}\right]$$

(if m is odd).

5. and 6) Assume [(a and p) are even and r is odd)] and[(mp≤r<(m+1) p)]. From b), $\Rightarrow$

$$\left[a^r\right],\left[(pa),a^{r-p}\right],\left[(pa)^2,a^{r-2p}\right],\cdots,\left[(pa)^m,a^{r-mp}\right]$$

are solutions of $x^p\in A(\beta)$ in S_n. Let

$$\lambda\in\left[a^r\right]\Rightarrow\lambda=\gamma_1\gamma_2\cdots\gamma_r\Rightarrow|\langle\gamma_i\rangle|=a$$

(even),

$(1\le i\le r)\Rightarrow\gamma_i$ is a product of an odd number similar to T_i of transpositions for all, $(1\le i\le k)\Rightarrow\gamma$ is a product of $(T_1+T_2+\ldots+T_r)$=(odd) number of transposetions, also for each $\lambda\in\left[(pa)^k,a^{r-kp}\right]$, where

$(1 \le k \le m) \Rightarrow \lambda = \lambda\beta \Rightarrow \beta = \beta_1\beta_2...\beta_{r-kp}$, where $|\langle\beta_i\rangle| = a$ (even), (1≤i≤r-kp) and $\gamma = \gamma_1\gamma_2...\gamma_k$, where

$|\langle\gamma_i\rangle| = pa$ (even) $(1 \le j \le k) \Rightarrow \beta_i$, is product of an odd number similar to L_i of transpositions for each(1≤i≤r-kp), and γ_j is product of an odd number similar to T_j of transpositions for each (1≤j≤k). If k is odd, then λ is a product of $(L_1+L_2+\ldots+L_{r-kp}) + (T_1+T_2+\ldots+T_k)$= (odd) + (odd) = (even) number of transpositions $\Rightarrow \lambda \in A_n$. If k is even, then λ is a product of $(L_1+L_2+\ldots+L_{r-kp}) + (T_1+T_2+\ldots+T_k)$ = (odd) + (even) = (odd) number of transpositions $\Rightarrow \lambda \notin A_n$. Then, if [(a and p) are even and r is odd)] and (mp≤r<(m+1)p), then the solutions of $x^p \in A(\beta)$ in A_n are

$$\left[(pa), a^{r-p}\right], \left[(pa)^3, a^{r-3p}\right], \cdots, \left[(pa)^m, a^{r-mp}\right]$$

(if m is odd), or

$$\left[(pa), a^{r-p}\right], \left[(pa)^3, a^{r-3p}\right], \cdots, \left[(pa)^{(m-1)}, a^{r-(m-1)p}\right]$$

(if m is even).

7) Assume (a is even and (p and r) are odd). For each

$\lambda \in [a^r]$ or $\lambda \in [a^r]$ or $\lambda \in \left[(pa)^k, a^{r-kp}\right]$, $(1 \le k \le m) \Rightarrow \lambda \notin A_n$,

then there is no solution of $x^p \in A(\beta)$ in A_n.

Remarks

Let p and q be different prime numbers and $\left[a^r\right]$ a conjugate class of symmetric group. If p|a, p|r and q does not divide a we defined collection of sets of conjugate classes of S_n as following:

- $W=\left\{\left[(pa)^i,(pqa)^j\right]\middle| i\right.$ and j are non-negative and solutions of the equation $\left. i+qj=-\frac{r}{p}\right\}$.
- $W_1=\left\{\pi\in W \middle| i+j \text{ is even}\right\}$.
- $W_2=\left\{\pi\in W \middle| (i \text{ and } j \text{ are even})\right.$

or (is odd and j is even)}

- $W_3=\left\{\pi\in W \middle| i+j \text{ is odd}\right\}$.
- $W_4=\left\{\pi\in W \middle| (i \text{ and } j \text{ are odd})\right.$

or (i is even and j is odd)}

*We note that $W=W_1\bigcup W_3$ & $W=W_2\bigcup W_4$

** $W_1\bigcap W_2=\left\{\pi\in W \middle| i \text{ and } j \text{ are even}\right\}$

$W_2\bigcup W_4=\phi$, $W_1\bigcup W_3=\phi$.

$W_1\bigcap W_4=\left\{\pi\in W \middle| (i \text{ and } j \text{ are odd})\right\}$,

$W_2\bigcap W_3=\left\{\pi\in W \mid i \text{ is odd and } j \text{ is even}\right\}$.

$W_3\bigcap W_4=\left\{\pi\in W \mid i \text{ is even and } j \text{ is odd}\right\}$.

Remarks

- If a, p and q are odd, then for each $\mu \in \pi$, where $\pi \in W$ we have μ is even.
- If a is even, then for each $\mu \in \pi$, where $\pi \in W$ we have (μ is even if $\pi \in W_1$) and (μ is odd if $\pi \in W_3$).
- If p is even, then for each $\mu \in \pi$, where $\pi \in W$ we have (μ is even if $\pi \in W_1$) and (μ is odd if $\pi \in W_3$).
- If q is even and a, p are odd, then for each $\mu \in \pi$, where $\pi \in W$ we have (μ is even if $\pi \in W_2$) and (μ is odd if $\pi \in$).

Theorem

Let A (β) be a conjugacy class of β in A_n, and $\beta \in \left[a^r\right] \cap H_n^c$, where $\left[a^r\right]$ is a class of S_n. If p and q are different two prime numbers and p|a and q does not divide a, then the solutions of $x^{pq} \in A(\beta)$ in A_n are:

- W if p|r and (a, p and q are odd).
- W_1 if p|r and (a or p is even).
- W_2 if p|r and (q is even & (a and p) are odd).
- Not exist if p does not divide r.
- Not exist if p|r, (a or p is even) and $W=W_3$.
- Not exist if p|r, (q is even & (a and p) are odd) and $W=W_4$.

Proof

Since $\beta \in \left[a^r\right] \cap H_n^c \cap A_n$, $A(\beta) = \left[a^r\right]$ and by (1.8) we have that the solution of $x^{pq} \in A(\beta)$ in S_n is:

a. W if p|r.
b. Not exist if p does not divide r.

- Assume p|r and (a, p and q are odd). Then from a) we have W is the solution set of $x^{pq} \in A(\beta)$ in S_n. Let $\mu \in \pi$ for each $\pi \in W$, we have μ is even permutation Then the solution set in A_n is W.

- Assume p|r and (a or p is even). Then from a) we have W is the solution set of $x^{pq} \in A(\beta)$ in S_n. Let $\mu \in \pi$ for each $\pi \in W$, we have (μ is even permutation, if $\pi \in W_1$) and (μ is odd permutation, if $\pi \in W_3$). Then the solution set in A_n is W_1.
- Assume p|r and (q is even & (a and p) are odd). Then from a) we have W is the solution set of $x^{pq} \in A(\beta)$ in S_n. Let $\mu \in \pi$ for each $\pi \in W$, we have (μ is even permutation, if $\pi \in W_2$) and (μ is odd permutation, if $\pi \in W_4$). Then the solution set in A_n is W_2.
- Assume p does not divide r. Then from b) we have no solution of $x^{pq} \in A(\beta)$ in $S_n \Rightarrow$ no solution of $x^{pq} \in A(\beta)$ in A_n.
- and 6) it is clear if p|r, (a or p is even) and W=W_3, then $W_1 = \phi$ and there exists no solution in A_n, also if p|r, (q is even & (a and p) are odd) and W=W_4, then $W_2 = \phi$ and there exists no solution in A_n.

Remarks

Let p and q be two different prime numbers and $\left[a^r\right]$ a conjugate class of symmetric group S_n, p does not divide a and q does not divide a we defined a collection of sets of conjugate classes of S_n as following:

$$D = \left\{\left[a^i, (pa)^j, (qa)^k, (pqa)^l\right] \middle| i, j, k \text{ and } l\right.$$

are non-negative integers and satisfying the equation i+jp+kp+lpq=r}.

$$D_1 = \left\{\pi \in D \middle| i, j, k \text{ and } l \text{ are all even or all odd}\right\}.$$

$$D_2 = \left\{\pi \in D \middle| i + j \text{ is even, } k + i \text{ is odd and } l + j \text{ is odd}\right\}$$

$$D_3 = \left\{\pi \in D \middle| i \text{ is odd, } j \text{ is even and } k + l \text{ is odd}\right\}.$$

$$D_4 = \left\{\pi \in D \middle| i \text{ is even, } j \text{ is odd and } k + l \text{ is odd}\right\}.$$

$$D_5 = \left\{\pi \in D \middle| j + l \text{ is even}\right\}.$$

$D_6 = \{\pi \in D | k+l \text{ is even}\}$.

$Q_1 = \{\pi \in D | (i \text{ and } j \text{ are even}) \text{and} (k+l \text{ is odd})\}$.

$Q_2 = \{\pi \in D | (i \text{ is even and } j \text{ is odd}) \text{and} (k+l \text{ is even})\}$.

$Q_3 = \{\pi \in D | (i \text{ is odd and } j \text{ is even}) \text{and} (k+l \text{ is even})\}$.

$Q_4 = \{\pi \in D | (i \text{ and } j \text{ are odd}) \text{and} (k+l \text{ is odd})\}$.

$Q_5 = \{\pi \in D | j+l \text{ is odd}\}$.

$Q_6 = \{\pi \in D | k+l \text{ is odd}\}$.

We can denote D as the following:

- $D=D_1 \cup D_2 \cup D_3 \cup D_4 \cup Q_1 \cup Q_2 \cup Q_3 \cup Q_4$
- $D=D_5 \cup Q_5$
- $D=D_6 \cup Q_6$

Remarks

- If a, p and q are odd, then for each $\mu \in \pi$, where $\pi \in D$, μ is even.
- If a is even, then for each $\mu \in \pi$, μ is even if $\pi \in D_1 \cup D_2 \cup D_3 \cup D_4$ and μ is odd if $\pi \in Q_1 \cup Q_2 \cup Q_3 \cup Q_4$.
- If p is even and (a and q) are odd, then for each $\mu \in \pi$, μ is even if $\pi \in D_5$ and μ is odd if $\pi \in Q_5$.
- If q is even and (a and p) are odd, then for each $\mu \in \pi$, μ is even if $\pi \in D_6$ and μ is odd if $\pi \in Q_6$.

Theorem

Let A(β) be the conjugacy class of β in A_n, and $\beta \in [a^r] \cap H_n^c$, where $[a^r]$ is a class of S_n, p and q are different two prime numbers. If p does not divide a and q does not divide a, then the solution of $x^{pq} \in A(\beta)$ in A_n is

- D, if a, p and q are odd
- T, where $T = D_1 \cup D_2 \cup D_3 \cup D_4$, if a is even
- D_5, if a and q are odd, and p is even
- D_6, if a and p are odd, and q is even
- does not exist, if a is even, and $D = Q_1 \cup Q_2 \cup Q_3 \cup Q_4$
- does not exist, if p is even, a and q are odd, and $D=Q_5$, or
- does not exist, if q is even, a and p are odd, and $D=Q_6$.

Proof

Considering that $\beta \in [a^r] \cap H_n^c \cap A_n$, $A(\beta) = [a^2]$ then by (2.2.11), D is the solution set of $x^{pq} \in A(\beta)$ in S_n.

- Assume a, p and q are odd. Let $\mu \in \pi$ for each $\pi \in D \Rightarrow \mu$ is even $\Rightarrow \mu \in A_n$. Then the solution set in A_n is D if a, p and q are odd.
- Assume a is even. Let $\mu \in \pi$ for each $\pi \in D \Rightarrow \mu$ is even $\Rightarrow \mu \in A_n$. Then the solution set in A_n is T, if a is even.
- Assume a and q are odd, and p is even. Let $\mu \in \pi$ for each $\pi \in D \Rightarrow \mu$ is even $\Rightarrow \mu \in A_n$. Then the solution set in A_n is D_5, if a and q are odd, and p is even.
- Assume a and p are odd, and q is even. Let $\mu \in \pi$ for each $\pi \in D \Rightarrow \mu$ is even $\Rightarrow \mu \in A_n$. Then the solution set in A_n is D_6, if a and p are odd, and q is even.
- Assume a is even and $D = Q_1 \cup Q_2 \cup Q_3 \cup Q_4 \Rightarrow .T = \phi$. Then there is no solution in A_n.

— and 7) Assume a and q are odd, p is even, and $D = Q_5 \Rightarrow D_5 = \phi$. Then there is no solution in A_n, also if a and p are odd, q is even, and $D = Q_6 \Rightarrow D_6 = \phi$. Then there is no solution in A_n.

The Number of the Solutions

Assume $\left\{\{\lambda_t\}_{t=1}^{T}, \{\gamma_j\}_{j=1}^{k}, \beta\right\}$ are even permutations where β and $\lambda_t \in H_n^c$ and $\lambda_j \in H_n$ for all $1 \leq t \leq T$ and $1 \leq j \leq k$. Then we can find the number of the solutions of class equation $x^d = \beta$ in A_n, where d is a positive integer number as follow:

— If $\left\{C^a(\lambda_t)\right\}_{t=1}^{T} \cup \left\{C^a(\lambda_j), C^{a-}(\gamma_j)\right\}_{j=1}^{k}$ are the solutions, then the number of solutions set is

$$\sum_{t=1}^{T} \frac{n!}{Z_{\alpha(\lambda_t)}} + \sum_{j=1}^{k} \frac{n!}{Z_{\alpha(\gamma_j)}},$$

— If $\left\{C^a(\lambda_t)\right\}_{t=1}^{T}$ are the solutions, then the number of solutions set is

$$\sum_{t=1}^{T} \frac{n!}{Z_{\alpha(\lambda_T)}},$$

— If $\left\{C^{a+}(\lambda_j), C^{a-}(\gamma_j)\right\}_{j=1}^{k}$ are the solutions, then the number of solutions set is

$$\sum_{j=1}^{k} \frac{n!}{Z_{\alpha(\gamma_j)}}.$$

Example

Find the solutions of $x^p \in A(\beta)$ in A_n, and the number of the solutions.

- If p = 3 and $\beta = (1\ 2)(3\ 4)(5\ 6)(7\ 8)$ in A_8.
- If p = 2 and $\beta = (1\ 2\ 3)(4\ 5\ 6)$ in A_6.

Solution

1. Since $\beta \in [2^4] \cap H_8^c$, a = 2, r = 4, p does not divide a, pm≤r<(1+m) p where, m = 1.
So a and r are even, and p is odd. Then by (3.1) the solutions of $x^3 \in A(\beta)$ in A_8 are $[2^4]$ and [2, 6], so the number of solutions is $\frac{(8)!}{2^4(4!)} + \frac{(8)!}{2 \times 6} = 3465$ permutations.

2 Since $\beta \in [3^2] \cap H_6^c$, a = 3, r = 2, p does not divide a, pm≤r<(1+m) p where, m = 1.
Also, since a is odd and p is even. Then by (3.1) the solution set of $x^2 \in A(\beta)$ in A_6 is $[3^2]$. So the number of solutions is $\frac{6!}{3^2 \times 2} = 40$ permutations.

Remark

If $C^\alpha(\lambda)$ conjugate class of λ in S_n belong to the solution set of class equation $x^d = \beta$ in A_n and $\lambda \in H_n$, then we denote to this set $C^\alpha(\lambda)$ by $C^\alpha(\lambda)^\pm$ or

$$\left\{C^\alpha(\lambda)^+, C^\alpha(\lambda)^-\right\}.$$

Example

Find the solution of $x^{35} \in A((1\ 2\ 5)(6\ 4\ 3)(8\ 10\ 15)(7\ 14\ 11)(13\ 9\ 12))$ in A_{15} and the number of the solutions.

Solution

Let p=7 and q=5, since $\beta = ((1\ 2\ 5)(6\ 4\ 3)(8\ 10\ 15)(7\ 14\ 11)(13\ 9\ 12))$ $\in [3^5] \cap H^c_{15}$

Then by (3.7) the solutions of $x^{35} \in A((1\ 2\ 5)(6\ 4\ 3)(8\ 10\ 15)(7\ 14\ 11)(13\ 9\ 12))$ in A^{15} are $[3^5]$, $[15]^+$ and$[15]^-$. So the number of the solutions set is

$$\frac{(15)!}{3^5(5)!} + \frac{(15)!}{15} = 87223136000 \text{ permutations.}$$

CONCLUDING REMARKS

By the Cayley's theorem: Every finite group G is isomorphic to a subgroup of the symmetric group S_n, for some $n \geq 1$. Then we can discuss these propositions. Let $x^d = g$ be class equation in finite group G and assume that $f : G \cong A_n$, for some $n \geq 1$and $f(g) \in H_n^c \cap C^\alpha$. The first question we are concerned with is: What is the possible value of d provided that there is no solution for $x^d = g$ in G? The second question we are concerned with is: what is the possible value of d provided that there is a solution for $x^d = g$ in G? And then we can find the solution and the number of the solution for $x^d = g$ in G by using Cayley's theorem and our theorems in this paper. In another direction, let G be a finite group, and $\pi_i(G) = \{g \in G | i$ the least positive integer number satisfy $g^i = 1\}$. If $|\pi_i(G)| = k_i$, then we write

$|\pi_i(G)| = \{g_{i1}, g_{i2}, \ldots, g_{ik_i}\}$ and $\prod = \{\pi_i(G)\}_{i \geq 1}$. For each

$g \in G$ and $g_{ij} \in \pi_i(G)$ we have $(gg_{ij}g^{-1})=1$. By the Cayley's theorem we can suppose that $(f : G \cong S_n)$ or $(f : G \cong A_n)$. Also the questions can be summarized as follows:

- Is $\Pi = \{\pi_i(G)\}_{i \geq 1}$ collection set of conjugacy classes of G?
- Is there some $i \geq 1$, such that $f^{-1}(C^\alpha) = \pi_i(G)$ for each C^α of A_n, where $(f : G \cong A_n)$?
- Is there some $i \geq 1$, such that $f^{-1}(C^\alpha) = \pi_i(G)$ for each C^α of S_n, where $(f : G \cong A_n)$?
- If $G \cong S_n$ and p(n) is the number of partitions of n, is $f^{-1}(C^\alpha) = \pi_i(G)|\Pi| = p(n)$?
- If $G \cong A_n$ and A_n has m ambivalent conjugacy classes. It is true that is also necessarily G has m ambivalent conjugacy classes?

Finally we will discuss if there is any relation between F_n, $\overline{F_n}$ and H_n in S_n and what is the possible value of d provided that there is a solution for $x^d=g$ in G where $f(g) \in H_n \cap C^\alpha$ and for some n to be:

- $n \in \theta$.
- $n \notin \theta$.

REFERENCES

1. H. Ishihara, H. Ochiai, Y. Takegahara and T. Yoshida, "p-divisibility of the number of solutions of $x^p = 1$ in a symmetric group," Annals of Combinatorics, Vol. 5, No. 2, 2001, pp. 197-210. doi:10.1007/PL00001300
2. N. Chigira, "The solutions of $x^d = 1$ in Finite Groups," Journal of Algebra, Vol. 180, No. 3, 1996, pp. 653-661. doi:10.1006/jabr.1996.0086
3. R. Brauer, "On a theorem of Frobenius," American Mathematical Monthly, Vol. 76, No. 1, 1969, pp. 12-15. doi:10.2307/2316779
4. Y. G. Berkovich, "On the number of elements of given order in finite p-group," Israel Journal of Mathematics, Vol. 73, No. 1, 1991, pp. 107-112. doi:10.1007/BF02773429

5. T. Ceccherini-Silberstein, F. Scarabotti and F. Tolli, "Representation Theory of the Symmetric Groups," Cambridge University Press, New York, 2010.
6. D. Zeindler, "Permutation matrices and the moments of Their Characteristic Polynomial," Electronic Journal of Probability, Vol. 15, No. 34, 2010, pp. 1092-1118.
7. J. J. Rotman, "An introduction to the theory of groups," 4th Edition, Springer-Verlag, New York, 1995.
8. G. D. James and A. Kerber, "The Representation Theory of the Symmetric Group," Addison-Wesley Publishing, Boston, 1984.
9. S. A. Taban, "Equations in Symmetric Groups," Ph.D. Thesis, University of Basra, Basra, 2007.
10. S. Mahmood and A. Rajah, "Solving the Class Equation $x^d = \beta$ in an Alternating Group for each $\beta \in H_n \cap C^a$ and $n \notin \theta$," Journal of the Association of Arab Universities for Basic and Applied Sciences, Vol. 10, No. 1, 2011, pp. 42-50. doi:10.1016/j.jaubas.2011.06.006

CITATION

1. S. Mahmood and A. Rajah, "Solving the Class Equation $x^d = \beta$ in an Alternating Group for Each $\beta \in C\alpha \cap H_n^c$ and $n > 1$," Advances in Linear Algebra & Matrix Theory, Vol. 2 No. 2, 2012, pp. 13-19. doi:10.4236/alamt.2012.22002.

Jordan Semi-Triple Multiplicative Maps on the Symmetric Matrices

Xiaoning Hao, Pengxiang Ren, and Runling An
Department of Mathematics, Taiyuan University of Technology, Staff Education and Training Center of Taiyuan Iron and Steel (Group) Co., Ltd., Taiyuan, China

ABSTRACT

In this paper, we show that if an injective map Φ on symmetric matrices $S_n(C)$ satisfies $\phi(ABC) = \phi(A)\phi(B)\phi(A), \forall A, B \in S_n(C)$ then $\phi(A) = \pm SA_f S^t$ for all $A \in S_n(C)$, where f is an injective homomorphism on C, Sis a complex orthogonal matrix and A_f is the image of A under f applied entry wise.

INTRODUCTION

It is an interesting problem to study the interrelation between the multiplicative and the additive structure of a ring or an algebra. Matindale in [1] proved that every multiplicative bijective map from a prime ring containing a nontrivial idempotent onto an arbitrary ring is additive. Thus, the multiplicative structure determines the ring structure for some rings. This result was utilized by P. Šemrl in [2] to describe the form of the semigroup isomorphisms of standard operator algebras on Banach spaces. Some other results on the additivity of multiplicative maps between operator algebras can be found in [3, 4]. Besides ring homomorphisms between rings, sometimes one has to consider Jordan ring homomorphisms. Note that, Jordan operator algebras have important applications in the mathematical foundations of quantum mechanics. So, it is also interesting to ask when the Jordan multiplicative structure determines the Jordan ring structure of Jordan rings or algebras.

Let R, R′ be two rings and let $\phi: R \to R'$ be a map. Recall that Φ is called a Jordan homomorphism if

$$\Phi(A+B)=\Phi(A)+\Phi(B),$$
$$\Phi(AB+BA)=\Phi(A)\Phi(B)+\Phi(B)\Phi(A)$$

for all $A,B \in R$. There are two basic forms of Jordan multiplicative maps, namely1) $\phi(ABA)=\phi(A)\phi(B)\phi(A)$ (Jordan semi-triple multiplicative map) for all $A,B \in R_2$ 2) $\phi\left(\frac{1}{2}AB+\frac{1}{2}BA\right)=\frac{1}{2}\phi(A)\phi(B)+\frac{1}{2}\phi(B)\phi(A)$ (Jordan multiplicative map) for all $A,B \in R$. It is clear that, if Φ is unital and additive, then these two forms of Jordan multiplicative maps are equivalent. But in general, for a unital map, we do not know whether they are still equivalent without the additivity assumption.

The question of when a Jordan multiplicative map is additive was investigated by several authors. Let $\phi\left(\frac{1}{2}AB+\frac{1}{2}BA\right)=\frac{1}{2}\phi(A)\phi(B)+\frac{1}{2}\phi(B)\phi(A)$ be a bijective map on a standard operator algebra. Molnár showed in [5] that if Φ satisfies

$$\phi(ABA)=\phi(A)\phi(B)\phi(A),$$

then $\phi\left(\frac{1}{2}AB+\frac{1}{2}BA\right)=\frac{1}{2}\phi(A)\phi(B)+\frac{1}{2}\phi(B)\phi(A)$ is additive. Later, Molnár in [5] and then Lu in [6] considered the cases that ϕ preserve the operation $\frac{1}{2}(AB+BA)$ and $AB+BA$, respectively, and proved that such ϕ is also additive. Thus, the Jordan multiplicative structure also determines the Jordan ring structure of the standard operator algebras. Later, in [7] we proved these Jordan multiplicative maps on the space of self adjoint operators space are Jordan ring isomorphism and thus are equivalent. In this paper, we consider the same question and give affirmative answer for the case of Jordan multiplicative maps on the Jordan algebras of all symmetric matrices. In fact, we study injective Jordan semi-triple multiplicative maps on the symmetric matrices $S_n(C)$, and show that such maps must be additive, and hence are Jordan ring homomorphisms.

Let us recall and fix some notations in this paper. Recall that $P \in S_n(C)$ is called an idempotent if $P^2 = P$. We define the order $\leq$ between idempotents as follows: $P \leq Q$ if and only if $PQ = QP = P$ for any idempotents. $P, Q \in S_n(C)$ For any $1 \leq j, k \leq n$, let E_{jk} be the matrix with 1 in the position (j,k) and zeros elsewhere, and I_j be the unit of $S_j(C)$.

MAIN RESULTS AND ITS PROOF

In this section, we study injective Jordan semi-triple multiplicative maps on $S_n(C)$, the following is the main result.

Theorem 2.1: An injective map $\Phi : S_n(C) \rightarrow S_n(C)$ is a Jordan semi-triple multiplicative map, that is

$$\Phi(ABA) = \Phi(A)\Phi(B)\Phi(A), \forall A, B \in S_n(\mathbf{C}) \tag{2.1}$$

If and only if there is an injective homomorphism f of C and a complex orthogonal matrix S such that $\Phi(A) = \pm SA_f S^t$ For all $A \in S_n(C)$.

Firstly, we give some properties of injective Jordan semi-triple multiplicative maps on $S_n(C)$.

Lemma 2.2: Let $\Phi : S_n(C) \to S_n(C)$ be an injective Jordan semi-triple multiplicative map. Then Φ sends idempotents to tripotents and moreover1) $\Phi(I_n)^2$ is an idempotent and

$$\Phi(A)$$
$$=\Phi(I_n)\Phi(A)\Phi(I_n)$$
$$=\Phi(I_n)^3\Phi(A)\Phi(I_n)$$
$$=\Phi(I_n)\Phi(A)\Phi(I_n)^3$$
$$=\Phi(I_n)^2\Phi(A)=\Phi(A)\Phi(I_n)^2$$

For all $A \in S_n(C)$, in particular

$$\Phi(I_n)^2\Phi(A)\Phi(I_n)^2=\Phi(A);$$

2) $\Phi(I_n)$ commutes with $\Phi(A)$ for every $A \in S_n(C)$;

3) $\Phi(P)^2 = \Phi(I_n)\Phi(P)$ Is an idempotent for each idempotent $P \in S_n(C)$;

4) A map $\psi : S_n(C) \to S_n(C)$ defined by $\psi(A) = \Phi(I_n)\Phi(A)$

For all $A \in S_n(C)$, is a Jordan semi-triple multiplicative map, which is injective if and only if Φ is injective.

For Ψ defined in Lemma 2.2, we can see that $\psi(P)^2 = \psi(P)$, and $P \le Q \Rightarrow \psi(P) \le \psi(Q)$ for any idempotents $P, Q \in S_n(C)$. Therefore, we have Corollary 2.3. Let $n, m \in N$ and $\Phi : S_n(C) \to S_m(C)$

Be an injective Jordan semi-triple multiplicative map. Then $m \ge n$. In the case, $m = n$ for each idempotent $P \in S_n(C)$ the rank of $\Psi(P)$ is equal to the rank of P. In particular,

$$\Psi(0)=\Phi(0)=0$$
$$\Psi(I)=\Phi(I)^2=I$$

And

$$\Psi\left(A^2\right)=\Psi\left(A\right)^2.$$

Now we give proof of Theorem 2.1. The main idea is to use the induction on n, the dimension of the matrix algebra, after proving the result for 2×2 matrices.

Proof of Theorem 2.1. In order to prove Theorem 2.1, it suffices to characterize Ψ. Note if

$$\Psi\left(A\right)=SA_fS^t$$

Then

$$\Phi\left(I_n\right)^2=\Psi\left(I_n\right)=I_n,$$

That is Φ (I_n) is invertible and

$$\Phi\left(I_n\right)=\Phi\left(I_n\right)^{-1}.$$

By Lemma 2.1, $\Phi(I_n)$ commutes with $\Phi(A)$ for all $A\in S_n(C)$. It follows that $\Phi(I_n)$ commutes with $\Psi(A)$ for all $A\in S_n(C)$. Therefore, if $\Psi(A)=SA_fS^t$, $\Phi(I_n)$ must be a scalar matrix. As $\Phi(I_n)\ =I_n$,.$\Phi(I_n)=\pm I_n$ and hence Φ has the desired form.

Therefore, we mainly characterize Ψ. The proofs are given in two steps.

Step 1. The proof for $S_2(C)$.

The matrix E_{11} is an idempotent of rank one. By Corollary 2.3, $\Psi(E_{11})$ is a rank one idempotent? It is well known that every idempotent matrix in $S_n(C)$ can be diagonalizable by complex orthogonal matrix. Thus, there exists a 2×2 orthogonal matrix S such that

$$\Psi(E_{11}) = SE_{11}S^t.$$

without loss of generality, we may assume that

$$\Psi(E_{11}) = E_{11}.$$

by Corollary 2.3 and from the following fact

$$\Psi\left((E_{12}+E_{21})^2\right) = I_2 = \Psi(E_{12}+E_{21})^2$$

and

$$\begin{aligned} &E_{11}\Psi(E_{12}+E_{21})E_{11} \\ &= \Psi(E_{11}(E_{12}+E_{21})E_{11}) = \Psi(0) = 0 \end{aligned}$$

we conclude that

$$\Psi|(E_{12}+E_{21}) = \begin{pmatrix} 0 & 1 \\ 1 & 0 \end{pmatrix}$$

or

$$\Psi(E_{12}+E_{21}) = \begin{pmatrix} 0 & -1 \\ -1 & 0 \end{pmatrix}.$$

Let, $V = \begin{pmatrix} 1 & 0 \\ 0 & -1 \end{pmatrix}$ by replacing Ψ with $V^t\Psi(.)V$ if necessary, we may assume that

$$\Psi(E_{12}+E_{21}) = E_{12}+E_{21}.$$

For E_{22}, since $\Psi(E_{22})$ is a rank one idempotent and satisfying $\Psi(E_{11})\Psi(E_{22})\Psi(E_{11})=0$ and

$$\Psi(E_{12}+E_{21})\Psi(E_{22})\Psi(E_{12}+E_{21})=\Psi(E_{11})=E_{11},$$

we have $\Psi(E_{22})=E_{22}$. Now for any

$$A=(a_{ij})\in S_2(\mathbf{C}),$$

let $B=(b_{ij})=\Psi(A)$. Then

$$\begin{aligned} b_{ii}E_{ii} &= E_{ii}BE_{ii}=\Psi(E_{ii})\Psi(A)\Psi(E_{ii}) \\ &= \Psi(E_{ii}AE_{ii})=\Psi(a_{ii}E_{ii}). \end{aligned}$$

Thus, the $(i,i)th$ entry of $\Psi(A)$ depends on the $(i,i)th$ entry of A only. Therefore, there exist injective functionals $f,g,:C\to C.$ such that f,g satisfy respectively $f(a^2b)=f(a)^2 f(b)$ and $g(a^2b)=g(a)^2 g(b)$ and

$$\Psi\begin{pmatrix} a_{11} & a_{12} \\ a_{12} & a_{22} \end{pmatrix}=\begin{pmatrix} f(a_{11}) & b_{12} \\ b_{12} & g(a_{22}) \end{pmatrix}.$$

From $f(1)=g(1)=1$, it is easy to verify that f,g is multiplicative. Next we prove that $f=g$. Let

$$A=\begin{pmatrix} \frac{1}{2} & \frac{1}{2} \\ \frac{1}{2} & \frac{1}{2} \end{pmatrix},$$ Since $(E_{12}+E_{21})A(E_{12}+E_{21})=A$

A and $A^2 = A$, we have

$$(E_{12}+E_{21})\Psi(A)(E_{12}+E_{21})$$
$$=\Psi((E_{12}+E_{21})A(E_{12}+E_{21}))=\Psi(A)$$

And $\Psi(A)^2 = \Psi(A)^2 = \Psi(A)$, hence $\Psi(A) = A$ or

$\Psi A = \begin{pmatrix} \frac{1}{2} & -\frac{1}{2} \\ -\frac{1}{2} & \frac{1}{2} \end{pmatrix}$ With $b^2 = \frac{1}{4}$.

Thus, $f\left(\frac{1}{2}\right) = g\left(\frac{1}{2}\right) = \frac{1}{2}$ and $f(2) = g(2) = 2$ since

f, g Is multiplicative. Let $j = \begin{pmatrix} 1 & 1 \\ 1 & 1 \end{pmatrix}$, then

$\Psi(j) = \begin{pmatrix} 1 & a \\ a & 1 \end{pmatrix}$. Note that $J(E_{12}+E_{21})J = 2J$, and

$$\Phi(J)(E_{12}+E_{21})\Phi(J) = \Phi(J(E_{12}+E_{21})J) = \begin{pmatrix} 2 & b \\ b & 2 \end{pmatrix},$$

That is

$$\begin{pmatrix} 1 & a \\ a & 1 \end{pmatrix}\begin{pmatrix} 0 & 1 \\ 1 & 0 \end{pmatrix}\begin{pmatrix} 1 & a \\ a & 1 \end{pmatrix} = \begin{pmatrix} 2 & b \\ b & 2 \end{pmatrix}.$$

This implies a=1 and $\Phi(J)=J$. Now by the fact AJA=J and $\Phi(A)J\Phi(A)=J$, we get $\Psi(A)=A$. For any $a\in C$, since

$$\begin{aligned}&f(a)J\\&=J(f(a)E_{11})J=\Psi(J)\Psi(aE_{11})\Psi(J)\\&=\Psi(aJE_{11}J)=\Psi(aJ)=\begin{pmatrix}f(a)&b\\b&g(a)\end{pmatrix},\end{aligned}$$

Thus $f(a)=g(a)$.

Next we prove that f is additive. Since

$$\Psi\left(\begin{pmatrix}a&0\\0&b\end{pmatrix}^2\right)=\Psi\begin{pmatrix}a&0\\0&b\end{pmatrix}^2,$$ And thus we have

$$\Psi\begin{pmatrix}a&0\\0&b\end{pmatrix}=\begin{pmatrix}f(a)&0\\0&f(b)\end{pmatrix}$$

For any $a,b\in C$. Moreover by the fact $JAJ=(a+b)J$, one can get that

$$\begin{aligned}&(f(a)+f(b))J\\&=J\Psi(A)J=\Psi(JAJ)\\&=\Psi((a+b)J)=\begin{pmatrix}f(a+b)&c\\\overline{c}&f(a+b)\end{pmatrix}\end{aligned}$$

and $f(a+b)=f(a)+f(b)$.

Finally, we prove

$$\Psi\begin{pmatrix} a & b \\ b & c \end{pmatrix} = \begin{pmatrix} f(a) & f(b) \\ f(b) & f(c) \end{pmatrix}$$

for any $a, b, c \in C$. Let

$$\Psi\begin{pmatrix} a & b \\ b & c \end{pmatrix} = \begin{pmatrix} f(a) & d \\ d & f(c) \end{pmatrix}.$$

By the fact that $J\begin{pmatrix} a & b \\ b & c \end{pmatrix}J = (a+2b+c)J$ and

$$J\Psi\begin{pmatrix} a & b \\ b & c \end{pmatrix}J$$

$$= \Psi(J)\Psi\begin{pmatrix} a & b \\ b & c \end{pmatrix}\Psi(J) = \Psi((a+2b+c)J),$$

We get $d = f(b)$ and $\Psi\begin{pmatrix} a & b \\ b & c \end{pmatrix} = \begin{pmatrix} f(a) & f(b) \\ f(b) & f(c) \end{pmatrix}$ for any $a, b, c \in C$.

Step 2: The induction. Let

$$P = I_{n-1} \oplus [0],$$

Then P is a rank n-1 idempotent, so is $\Psi(P)$ by Corollary 2.3. Therefore, there exists an orthogonal matrix S such that $\Psi(P) = SPS^t$. Replacing Ψ by the map $A \mapsto S^t\Psi(A)S$, we may assume that $\Psi(P) = P$.

For any $\hat{A} \in S_{n-1}(C)$, let $A = \hat{A} \oplus [0]$. Then PAP=A implies

$$P\Psi(A)P=\Psi(P)\Psi(A)\Psi(P)=\Psi(PAP)=\Psi(A).$$

It follows that $\Psi\left(\hat{A}\oplus[0]\right)=\Psi(A)=\hat{X}\oplus[0]$ for some matrix $\hat{X}\in S_{n-1}(C)$. Define the map $\hat{\Psi}$ on $S_{n-1}(C)$

By $\hat{\Psi}(\hat{A})=\hat{X}$. It is easy to check that $\hat{\Psi}$ is an injective Jordan semi-triple multiplicative map on $S_{n-1}(C)$. Furthermore, $\Psi(P)=P$ implies that $\hat{\Psi}(I_{n-1})=I_{n-1}$. By the induction hypothesis there is a $(n-1)\times(n-1)$ orthogonal matrix $\hat{S}$ and an injective homomorphism f on C such that $\hat{\Psi}(\hat{A})=\pm\hat{S}\hat{A}_f\hat{S}^t$.

Let S be the matrix $\hat{S}\oplus[1]$. Without loss of generality, we assume that $\hat{\Psi}(\hat{A})=\hat{A}_f$ for all $\hat{A}\in S_{n-1}(C)$. This is equivalent to $\Psi\left(\hat{A}\oplus[0]\right)=\hat{A}_f\oplus[0]$. For any

$$A=\begin{pmatrix} A_{11} & x \\ x^t & a_{22} \end{pmatrix}\in S_n(C) \text{ with } A_{11}\in S_{n-1}(C), x\in C^{n-1}$$

and $a_{22}\in C$, we have $PAP=A_{11}\oplus[0]$.

Thus,

$$\begin{aligned} &P\Psi(A)P \\ &=\Psi(P)\Psi(A)\Psi(P)=\Psi(0)=(A_{11})_f\oplus[0] \end{aligned} \qquad (*)$$

Let us define matrices R$_i$ for each $i\in\{1,2,\cdots,n-1\}$ by

$$R_i=I_n-E_{ii}-E_{nn}+E_{in}+E_{ni}.$$

For an arbitrary i, from (*) we have

$$P\Psi(R_i)P = (I_{n-1} - E_{ii}) \oplus [0].$$

Then there exists $x \in C^{n-1}$ and $y \in C$ such that

$$\Psi(R_i) = \begin{pmatrix} I_{n-1} - E_{ii} & x \\ x^t & y \end{pmatrix}.$$

From the equality $\Psi(R_i)^2 = \Psi(R_i^2) = \Psi(I_n) = I_n$, we get that $I_{n-1} - E_{ii} + xx^i = I_{n-1}$ and $x'x + y^2 = 1$. These equality implies that $xx^t = E_{ii}$ and

$$y^2 = 1 - x^t x = 1 - tr(xx^t) = 1 - tr(E_{ii}) = 0.$$

Hence only the ith entries a_i of x are nonzero and $a_i^2 = 1$. It follows that

$$\Psi(R_i) = I_n - E_{ii} - E_{nn} + a_i E_{in} + a_i E_{ni}.$$

Next, take any two distinct $i, j \in \{1, 2, \cdots, n-1\}$. From

$$R_i R_j R_i = I_n - E_{ii} - E_{jj} + E_{ij} + E_{ji}$$

And using (*), we get

$$\Psi(I_n - E_{ii} - E_{jj} + E_{ij} + E_{ji})$$
$$= \Psi(R_i)\Psi(R_j)\Psi(R_i)$$
$$= I_n - E_{ii} - E_{jj} + a_i a_j E_{ij} + a_i a_j E_{ji},$$

Which implies that $a_i a_j = 1$. Let $D = 1_{n-1} \oplus [-1]$, then $D\Psi(R_1)D^t = R_1$, so we may assume that $\Psi(R_1) = R_1$. Furthermore by the equality

$$R_1 R_i R_1 = I_n - E_{11} - E_{ii} + E_{1i} + E_{i1}$$

and $R_1 \Psi(R_1) R_1 = \Psi(R_1 R_i R_1)$, we obtain $\Psi(R_i) = R_i$

Next we prove that $\Psi(A) = A_f$ for any $A \in S_n(C)$.

Let us fix some $i \in \{1, 2, \cdots, n-1\}$. As $n > 2$, there is another $j \in \{1, 2, \cdots, n-1\}$ such that

$$aE_{ni} + aE_{in} = R_j \left(aE_{ij} + aE_{ji} \right) R_j.$$

Then for any $a, b \in C$,

$$\begin{aligned}
&\Psi(aE_{nn}) \\
&= \Psi(R_1)\Psi(aE_{11})\Psi(R_1) \\
&= R_1 f(a) E_{11} R_1 = f(a) E_{nn}
\end{aligned}$$

and

$$\begin{aligned}
&\Psi(bE_{ni} + bE_{in}) \\
&= \Psi\left(R_j \left(bE_{ij} + bE_{ji}\right) R_j\right) \\
&= \Psi\left(R_j\right)\Psi\left(bE_{ij} + \bar{b}E_{ji}\right)\Psi\left(R_j\right). \\
&= R_j \left(f(b) E_{ij} + f(b) E_{ij}\right) R_j \\
&= f(b) E_{ni} + f(b) E_{in}
\end{aligned}$$

Thus, for any

$$A = \begin{pmatrix} aE_{ii} & x \\ x^t & a \end{pmatrix}$$

Where $x \in C^{n-1}$ has only one nonzero entry in the ith position, we have $\Psi(A) = A$. For any $A \in S_n(C)$, let
And

$$A = \begin{pmatrix} \hat{A} & x \\ \overline{x}^t & a \end{pmatrix}.$$

From $E_{nn}AE_{nn} = aE_{nn}$, we have

$$\begin{aligned} &bE_{nn} \\ &= E_{nn}\Psi(A)E_{nn} = \Psi(E_{nn})\Psi(A)\Psi(E_{nn}) \\ &= \Psi(E_{nn}AE_{nn}) = f(a)E_{nn} \end{aligned}$$

And $f(a) = b$. For any $i \in \{1, 2, \cdots, n-1\}$, since

$$(E_{in} + E_{ni})A(E_{in} + E_{ni}) = B = \begin{pmatrix} \hat{B} & z \\ z^t & \alpha \end{pmatrix}$$

Where $\hat{B} \in S_{n-1}(C)$ and Z have only one nonzero entry a_{nn} and a_{in} in the $(i,i)th$ and ith position respectively, α is equal to the $(i,i)th$ entry of A, thus we have

$$\begin{aligned}&(E_{in}+E_{ni})\Psi(A)(E_{in}+E_{ni})\\&=\Psi(E_{in}+E_{ni})\Psi(A)\Psi(E_{in}+E_{ni})\\&=\Psi((E_{in}+E_{ni})A(E_{in}+E_{ni}))\end{aligned}$$

and so $y = x_f$. The proofs are complete.

By Theorem 2.1, we can characterize another two forms of Jordan multiplicative maps on $S_n(C)$.

Theorem 2.4. An injective map

$$\Phi : S_n(\mathbf{C}) \to S_n(\mathbf{C})$$

Satisfies

$$\begin{aligned}&\Phi\left(\frac{1}{2}(ABC+CBA)\right)\\&=\frac{1}{2}\Phi(A)\Phi(B)\Phi(C)+\frac{1}{2}\Phi(C)\Phi(B)\Phi(A)\\&\forall A,B,C\in S_n(\mathbf{C})\end{aligned} \tag{2.2}$$

If and only if there is an injective homomorphism f on C and a complex orthogonal matrix S such that $\Phi(A) = \pm SA_fS^t$ For all $A \in S_n(C)$.

Proof. Let $C = A$ in Equation (2.2), we get

$$\Phi(ABA) = \Phi(A)\Phi(B)\Phi(A),$$

That is, Φ is a Jordan semi-triple multiplicative map? Consequently, Φ has the desired form by Theorem 2.1.

Since every ring homomorphism on R is an identity map, thus by Theorem 2.1, Theorem 2.4, we get Corollary 2.5. Let $\Phi : S_n(R) \to S_n(R)$ be an injective map. Then the following condition are equivalent

1. $\Phi(ABA) = \Phi(A)\Phi(B)\Phi(A), \forall A, B \in S_n(\mathbf{R})$

   $\Phi\left(\frac{1}{2}(ABC + CBA)\right)$

2. $= \frac{1}{2}\Phi(A)\Phi(B)\Phi(C) + \frac{1}{2}\Phi(C)\Phi(B)\Phi(A)$

   $\forall A, B, C \in S_n(\mathbf{R}),$

3. There is a real orthogonal matrix S such that $\Phi(A) = \pm SAS^t$ For all $A \in S_n(R)$.

At the end of this section, we characterize bijective maps on $S_n(C)$ preserving $ABC + CBA$.

Theorem 2.6: A bijective map $\Phi : S_n(C) \to S_n(C)$ satisfies

$$\Phi(ABC + CBA)$$
$$= \Phi(A)\Phi(B)\Phi(C) + \Phi C\Phi(B)\Phi(A) \tag{2.3}$$
$$\forall A, B, C \in S_n(\mathbf{C}),$$

If and only if there is a ring isomorphism f on C and a complex orthogonal matrix S such that $\Phi(A) = \pm SA_f S^t$ For all $A \in S_n(C)$.

Proof: It is enough to check the "only if" part. Letting $A = B$ in Equation (2.3), we get

$$\Phi(A^2C + CA^2) = \Phi(A)^2\Phi(C) + \Phi(C)\Phi(A)^2.$$

Taking $A=\frac{1}{\sqrt{2}}$ and $\Phi(C)=I$, we get $I=2\Phi\left(\frac{I}{\sqrt{2}}\right)^2$, and thus

$$\Phi\left(\frac{I}{\sqrt{2}}\right)^2=\frac{I}{2} \tag{2.4}$$

Letting $A=C$ in Equation (2.3), we get

$$\Phi(2ABA)=2\Phi(A)\Phi(B)\Phi(A).$$

Taking $A=\frac{1}{\sqrt{2}}$, we get

$$\Phi(B)=2\Phi\left(\frac{I}{\sqrt{2}}\right)\Phi(B)\Phi\left(\frac{I}{\sqrt{2}}\right).$$

Multiplying this equality by $\Phi\left(\frac{I}{\sqrt{2}}\right)$ from the left side, by Equation (2.4) we get

$$\Phi\left(\frac{I}{\sqrt{2}}\right)\Phi(B)=\Phi(B)\Phi\left(\frac{I}{\sqrt{2}}\right)$$

For any $B\in S_n(C)$, and hence $\Phi\left(\frac{I}{\sqrt{2}}\right)=\lambda I$ for some scalar $\lambda\in C$. By Equation (2.4), we obtain

$$\Phi\left(\frac{I}{\sqrt{2}}\right)=\pm\frac{I}{\sqrt{2}}.$$

If $\Phi\left(\frac{1}{\sqrt{2}}\right)=\frac{I}{\sqrt{2}.}$, let $\Psi(A)=-\Phi(A)$, then Ψ also meets Equation (2.3) and $\Phi\left(\frac{I}{\sqrt{2}}\right)=\frac{I}{\sqrt{2}.}$. So without loss of generality, we assume $\Phi\left(\frac{I}{\sqrt{2}}\right)=\frac{I}{\sqrt{2}.}$. By letting C=A and $B=\frac{1}{\sqrt{2}}$ in Equation (2.3), we get $\Phi(\sqrt{2}A^2)=\sqrt{2}\Phi(A)^2$ And $\Phi(\sqrt{2}A^2)=2\Phi(A)^2$ for all $A\in S_n(C)$. Consequently

$$\Phi(I)=\Phi\left(2\left(\frac{I}{\sqrt{2}}\right)^2\right)=2\Phi\left(\frac{I}{\sqrt{2}}\right)^2=I.$$

Now letting $B=C=I$ in Equation (2.3) we get

$$\Phi(2A)=2\Phi(A).$$

Thus, $2\Phi(ABA)=\Phi(2ABA)=2\Phi(A)\Phi(B)\Phi(A)$

And $2\Phi(ABA)=\Phi(A)\Phi(B)\Phi(A)$ by taking A=C in Equation (2.3). Therefore, Φ has desired form by surjectivity of Φ and Theorem 2.1.

In particular, we have Corollary 2.7. A bijective map $\Phi:S_n(R)\rightarrow S_n(R)$ satisfies

$$\Phi(ABC+CBA)$$
$$=\Phi(A)\Phi(B)\Phi(C)+\Phi(C)\Phi(B)\Phi(A)$$
$$\forall A,B,C\in S_n(\mathbf{R}),$$

If and only if there is a real orthogonal matrix S such that $\Phi(A)=\pm SAS^t$ For all $A\in S_n(R)$.

Remark 2.8. We do not know whether the surjective assumption in Theorem 2.6 and Corollary 2.7 can be omitted.

REFERENCES

1. W. S. Matindale III, "When Are Multiplicative Mappings Additive?" Proceedings of the American Mathematical Society, Vol. 21, No. 3, 1969, pp. 695-698. doi:10.1090/S0002-9939-1969-0240129-7
2. P. Šemrl, "Isomorphisms of Standard Operator Algebras," Proceedings of the American Mathematical Society, Vol. 123, No. 6, 1995, pp. 1851-1855.
3. L. Molnár, "On Isomorphisms of Standard Operator Algebras," Studia Mathematica, Vol. 142, 2000, pp. 295–302.
4. J. Hakeda, "Additivity of *-Semigroup Isomorphisms among *-Algebra," Bulletin of the London Mathematical Society, Vol. 18, No. 1, 1986, pp. 51-56. doi:10.1112/blms/18.1.51
5. L. Molnár, "Jordan Maps on Standard Operator Algebras," In: Z. Daroczy and Z. Páles, Eds., Functional Equations-Results and Advances, Kulwer Academic Publishers.
6. F. Lu, "Additivity of Jordan Maps on Standard Operator Algebras," Linear Algebra and its Applications, Vol. 357, No. 1-3, 2002, pp. 123-131. doi:10.1016/S0024-3795(02)00367-1
7. R. L. An and J. C. Hou, "Additivity of Jordan Multiplicative Maps on Jordan Operator Algebras," Taiwanese journal of mathematics, Vol. 10, No. 1, 2006, pp. 45-64.

CITATION

1. X. Hao, P. Ren and R. An, "Jordan Semi-Triple Multiplicative Maps on the Symmetric Matrices," Advances in Linear Algebra & Matrix Theory, Vol. 3 No. 2, 2013, pp. 11-16. doi: 10.4236/alamt.2013.32003.

Generalized Form of Hermite Matrix Polynomials via the Hypergeometric Matrix Function

Raed S. Batahan
Department of Mathematics, Faculty of Science, Hadhramout University, Hadhranout, Yemen

9

ABSTRACT

The object of this paper is to present a new generalization of the Hermite matrix polynomials by means of the hypergeometric matrix function. An integral representation, differential recurrence relation and some other properties of these generalized forms are established here. Moreover, some new properties of the Hermite and Chebyshev matrix polynomials are obtained. In particular, the two-variable and two-index Chebyshev matrix polynomials of two matrices are presented.

INTRODUCTION

Special functions have been developed deeply in the last decades to special matrix functions due to their applications in certain areas of statistics, physics and engineering. The Laguerre and Hermite matrix polynomials are introduced in [1] as examples of right orthogonal matrix polynomial sequences for appropriate right matrix moment functionals of integral type. The Hermite matrix polynomials, $H_n(x, A)$ have been introduced and studied in [2] [3] where $H_n(x, A)$ involves a parameter $A \in C^{r \times r}$ whose eigenvalues are all situated in the open right-hand half of the complex plane. The two-variable Hermite matrix polynomials, $H_n(x,y,A)$ have been presented in [4] as an extension of. $H_n(x,A)$ Moreover, some properties and other generalizations of $H_n(x,A)$ are given in [5] –[11]. As one of qualitative properties of the two-variable Hermite matrix polynomials, the Chebyshev matrix polynomials of the second kind are introduced in [4] , see also [12] [13].

The main aim of this paper is to consider a new generalization of the Hermite matrix polynomials and to derive some properties for the Hermite and Chebyshev matrix polynomials. The structure of this paper is the following. This section summarizes previous results essential in the rest of the paper and gives the development of the two-variable Hermite matrix polynomials. A matrix version of Kummer's first formula for the confluent hypergeometric matrix function is derived in Section 2. In Section 3, the addition theorem and three terms recurrence relation for the Chebyshev matrix polynomials of the second kind are obtained and further we introduce and study the two-variable and two-index Chebyshev matrix polynomials of two matrices. Finally, Section 4 deals with the study of the Generalized Hermite matrix polynomials by means of the hypergeometric matrix function.

In what follows, $C^{r\times r}$ denotes the set of complex matrices of size $r\times r$ and the matrices I and θ in $C^{r\times r}$ denote the matrix identity and the zero matrix of order r, respectively. For a matrix A in $C^{r\times r}$, its spectrum $\sigma(A)$ denotes the set of all eigenvalues of A. We say that a matrix A in $C^{r\times r}$ is a positive stable if

$$\mathrm{Re}(\mu) > 0 \text{ for every eigenvalue } \mu \in \sigma(A). \tag{1}$$

If f (z) and g (z) are holomorphic functions of the complex variable z, which are defined in an open set Ω of the complex plane and A is a matrix in $C^{r\times r}$ with $\sigma(A) \subset \Omega$, then from the properties of the matrix functional calculus ([14], p. 558), it follows that $f(A)g(A) = g(A)f(A)$.

If D_0 is the complex plane cut along the negative real axis and Log (z) denotes the principle logarithm of z then $z^{1/2}$ represents $\exp\left(\frac{1}{2}\mathrm{Log}(z)\right)$.

If A is a matrix in $C^{r\times r}$ with $\sigma(A) \subset D_0$, then $A^{\frac{1}{2}} = \sqrt{A}$ denotes the image by $z^{1/2}$ of the matrix functional calculus acting on the matrix A.

Let A be a matrix in $C^{r\times r}$ which satisfies the condition (1). The two-variable Hermite matrix polynomials [2VHMPs] are generated by [4]

$$\exp\left(xt\sqrt{2A} - yt^2 I\right) = \sum_{n\geq 0} \frac{1}{n!} H_n(x,y,A)t^n, \tag{2}$$

and are defined by the series

$$H_n(x,y,A) = n! \sum_{k=0}^{\lfloor n/2 \rfloor} \frac{(-1)^k y^k}{k!(n-2k)!} \left(x\sqrt{2A}\right)^{n-2k}, \tag{3}$$

where $\lfloor v \rfloor$ is the standard floor function which maps a real number v to its next smallest integer.

It is therefore evident, for y=1, that

$$H_n(x,1,A) = H_n(x,A), \tag{4}$$

where $H_n(x,A)$ is the Hermite matrix polynomials as given in [2]. Furthermore, $H_n(x,y,A) = y^{n/2} H_n(x/\sqrt{y}, A)$.

According to [4], we have

$$\frac{\partial^k}{\partial x^k} H_n(x,y,A) = \left(\sqrt{2A}\right)^k \frac{n!}{(n-k)!} H_{n-k}(x,y,A);\ 0 \leq k \leq n, \tag{5}$$

$$\frac{\partial^k}{\partial y^k} H_n(x,y,A) = \frac{(-1)^k n!}{(n-2k)!} H_{n-2k}(x,y,A);\ 0 \leq k \leq \left[\frac{n}{2}\right]. \tag{6}$$

Also, the 2VHMPs appear as a solution of the second order matrix differential equation in the form

$$\left[y\frac{\partial^2}{\partial x^2}-xA\frac{\partial}{\partial x}+nA\right]H_n(x,y,A)=0. \tag{7}$$

And satisfy the three terms recurrence relationship

$$H_n(x,y,A)=x\sqrt{2A}H_{n-1}(x,y,A)-2(n-1)yH_{n-2}(x,y,A);\ \ n\geq 2 \tag{8}$$

with $H_0(x,y,A)=I$ and $H_1(x,y,A)=x\sqrt{2A}$.

From (5), the relation (8) gives

$$H_n(x,y,A)=\left(x\sqrt{2A}-2y\left(\sqrt{2A}\right)^{-1}\frac{\partial}{\partial x}\right)H_{n-1}(x,y,A). \tag{9}$$

Iteration (9) yields a another representation of the 2VHMPs in the form

$$H_n(x,y,A)=\left(x\sqrt{2A}-2y\left(\sqrt{2A}\right)^{-1}\frac{\partial}{\partial x}\right)^n(I). \tag{10}$$

Another remarkable representation of the 2VHMPs, which is due essentially to ([4], Theorem 7), has the elegant form:

$$H_n(x,y,A)=\exp\left(-y(2A)^{-1}\frac{\partial^2}{\partial x^2}\right)\left(x\sqrt{2A}\right)^n. \tag{11}$$

Applying (11) provides the formula

$$H_n(x,y+z,A)=\exp\left(-y(2A)^{-1}\frac{\partial^2}{\partial x^2}\right)H_n(x,z,A). \tag{12}$$

In fact, the addition and multiplication theorems are

$$H_n\left(ax+bz,y,A\right)=\sum_{k=0}^{n}\binom{n}{k}H_{n-k}\left(ax,\frac{y}{2},A\right)H_k\left(bz,\frac{y}{2},A\right), \tag{13}$$

and

$$a^n H_n\left(x,y,A\right)=H_n\left(ax,a^2y,A\right). \tag{14}$$

If A(k,n) and B(k,n) are matrices in $C^{r\times r}$ for $n \geq 0$ and $k \geq 0$, then it follows that [10] [15] [16]:

$$\sum_{n\geq 0}\sum_{k\geq 0}A(k,n)=\sum_{n\geq 0}\sum_{k=0}^{n}A(k,n-k), \tag{15}$$

and

$$\sum_{n\geq 0}\sum_{k=0}^{n}B(k,n)=\sum_{n\geq 0}\sum_{k=0}^{\lfloor n/2\rfloor}B(k,n-k). \tag{16}$$

THE CONFLUENT HYPERGEOMETRIC MATRIX FUNCTION

In this section, the confluent hypergeometric matrix function is given. For the sake of clarity in the presentation, we recall some concepts and results related to the generalized hypergeometric matrix functions that may be found in [15] [18] [19].

The reciprocal gamma function denoted by $\Gamma^{-1}(z)=1/\Gamma(z)$ is an entire function of the complex variable z. Then, for any matrix A in $C^{r\times r}$, the image of $\Gamma^{-1}(z)$ acting on A, denoted by $\Gamma^{-1}(A)$ is a well-defined matrix. Furthermore, if

$$A + nI \text{ is invertible for every integer } n \geq 0, \tag{17}$$

Then $\Gamma(A)$ is invertible, its inverse coincides with $\Gamma^{-1}(A)$ and it follows that ([17], p. 253)

$$(A)_n = A(A+I)\cdots(A+(n-1)I); \quad n \geq 1, \tag{18}$$

with $(A)_0 = I$.

If A is a positive stable matrix in $C^{r\times r}$, then the gamma matrix function, $\Gamma(A)$ is well defined as [18]

$$\Gamma(A) = \int_0^{\infty} e^{-t} t^{A-I} dt. \tag{19}$$

From ([19], p. 206), we have

$$(A)_n = \Gamma(A+nI)\Gamma^{-1}(A); \quad n \geq 0. \tag{20}$$

Definition 2.1 [15]: Let p and q be two non-negative integers. The generalized hypergeometric matrix function is defined in the form:

$${}_pF_q(A_1,\cdots,A_p;B_1\ldots,B_q;z) = \sum_{n\geq 0}(A_1)_n\cdots(A_p)_n\left[(B_1)_n\right]^{-1}\cdots\left[(B_q)_n\right]^{-1}\frac{z^n}{n!}, \tag{21}$$

Where A_i and $B_j (1 \leq i \leq p, 1 \leq j \leq q)$ are matrices in $C^{r\times r}$ such that the matrices B_j satisfy the condition (17).

According to [15], it follows that:

- If $p \leq q$, then the power series (21) converges for all finite.
- If $p = q+1$, then the power series (21) is absolutely convergent for $|z| < 1$ and diverges for $|z| > 1$.

— If $p \leq q$ then the power series (21) diverges for $z \neq 0$.

With p=1 and q=0 in (21), one gets the following relation due to ([3], p. 213)

$$(1-z)^{-A} = \sum_{n\geq 0} \frac{1}{n!}(A)_n\, z^n, \quad |z|<1, \tag{22}$$

Which can be written by (19) and (20) in the form

$$(1-z)^{-A} = \Gamma^{-1}(A)\int_0^{\infty} e^{(z-1)t} t^{A-I} \mathrm{d}t. \tag{23}$$

For p=2 and q=1 in (21), we obtain the hypergeometric matrix function as given in [3] in the form

$$F(A,B;C;z) = \sum_{n\geq 0} (A)_n (B)_n \left[(C)_n\right]^{-1} \frac{z^n}{n!}; \quad |z|<1. \tag{24}$$

Moreover, the confluent hypergeometric matrix function is well defined for all finite z, when p=q=1, in the form

$$\Phi(A;B;z) = \sum_{n\geq 0} (A)_n \left[(B)_n\right]^{-1} \frac{z^n}{n!}. \tag{25}$$

One can easily get the following result.

Proposition 2.2

$$\frac{\mathrm{d}}{\mathrm{d}z}\Phi(A;B;z) = A\Phi(A+I;B+I;z)B^{-1} \tag{26}$$

and

$$\frac{d^k}{dz^k}\Phi(A;B;z)=(A)_k\,\Phi(A+kI;B+kI;z)\left[(B)_k\right]^{-1}. \tag{27}$$

In [20], the following theorem was proved:

Theorem 2.3: Let A and B be two matrices in $C^{r\times r}$ such that

- A and B-A are positive stable
- AB=BA.
- B+jI is invertible for all ≥ 0.

Then for a positive integer $n \geq 0$ the following holds

$$F(-nI,A;B;1)=\Gamma(B-A+nI)\Gamma^{-1}(B+nI)\Gamma^{-1}(B-A)\Gamma(B). \tag{28}$$

Indeed, by (20) we can rewrite the formula (28) in the form

$$F(-nI,A;B;1)=(B-A)_n\left[(B)_n\right]^{-1}. \tag{29}$$

A matrix version of Kummer's first formula for the confluent hypergeometric matrix function is presented in the following theorem:

Theorem 2.4: Let A and B be two matrices in $C^{r\times r}$ satisfy the conditions of Theorem 2.3. Then

$$\Phi(B-A;B;-z)=e^{-z}\Phi(A;B;z). \tag{30}$$

Proof: From (15) and (25) we have

$$e^{-z}\Phi(A;B;z)=\sum_{n\geq 0}\sum_{k=0}^{n}\frac{(-nI)_k}{k!}(A)_n\left[(B)_n\right]^{-1}\frac{(-z)^n}{n!}. \tag{31}$$

By (29) and taking into account the conditions of Theorem 2.3 we find

$$e^{-z}\Phi(A;B;z)=\sum_{n\geq 0}(B-A)_n\left[(B)_n\right]^{-1}\frac{(-z)^n}{n!}, \tag{32}$$

and so (30) follows.

GENERALIZED CHEBYSHEV MATRIX POLYNOMIALS

In [20], the Chebyshev matrix polynomials of the first kind $T_n(x,A)$ was defined by

$$T_n(x,A)=\sum_{k=0}^{n}\frac{(-1)^k n(n+k-1)!}{2^k k!(n-k)!}\Gamma(A)\Gamma^{-1}(A+kI)(1-x)^k. \tag{33}$$

From (20) and (25) with the use of

$$n!=\int_0^\infty e^{-t}t^n dt, \tag{34}$$

We give an integral representation of $T_n(x,A)$ in the form

$$T_n(x,A)=\frac{1}{(n-1)!}\int_0^\infty e^{-t}t^{n-1}\Phi(-nI;A;t(1-x)/2)dt. \tag{35}$$

The generalized Chebyshev matrix polynomials of the second kind [GCMPs] are defined by the series [4]

$$U_n(x,y,A)=\sum_{k=0}^{[n/2]}\frac{(-1)^k(n-k)!\left(\sqrt{2A}\right)^{n-2k}}{k!(n-2k)!}x^{n-2k}y^k \tag{36}$$

And specified by the integral representation

$$U_n(x,y,A)=\frac{1}{n!}\int_0^\infty e^{-t}t^n H_n\left(x,\frac{y}{t},A\right)dt. \tag{37}$$

According to (14), the integral representation (37) becomes

$$U_n(x,y,A)=\frac{1}{n!}\int_0^{\infty}e^{-t}H_n(xt,yt,A)\,dt. \tag{38}$$

The use of the relations (5) and (8) in (37) yields the differential recurrence relation

$$nU_n(x,y,A)=x\frac{\partial}{\partial x}U_n(x,y,A)-2y\left(\sqrt{2A}\right)^{-1}\frac{\partial}{\partial x}U_{n-1}(x,y,A).$$

According to [4], we have $U_n(x,1,A)=U_n(x,A)$ and, $U_n(x,y,A)=y^{n/2}U_n(x/\sqrt{y},A)$ where $U_n(x/\sqrt{y},A)$ is the Chebyshev matrix polynomials of the second kind [CMPs].

As a direct consequent of ([4], Lemma 5), we state the following result.

Proposition 3.1: for a real number K>2, it follows that

$$\left\|U_n(x,y,A)\right\|\le K^n\sqrt{\frac{(2y)^n}{n!}}\left(1-\frac{x^2}{2}\right)^{-(n+2)/2}\Gamma\left(\frac{n+2}{2}\right), \tag{39}$$

Where $|x|<K\|\sqrt{2A}\|$ and y>0.

Let us now introduce the two-variable and two-index Chebyshev matrix polynomials of two matrices [2V2ICMP] through the integral representation

$$U_{m,n}(x,y,A,B)=\frac{1}{m!n!}\int_0^{\infty}e^{-t}H_m(xt,t,A)H_n(yt,t,B)\,dt, \tag{40}$$

Where A and B are two matrices in $C^{r\times r}$ satisfy the condition (1). From (3) and (34) we obtain that

$$U_{m,n}(x,y,A,B)=\sum_{i=0}^{\lfloor m/2\rfloor}\sum_{j=0}^{\lfloor n/2\rfloor}\frac{(-1)^{i+j}(m+n-i-j)!}{i!j!(m-2i)!(n-2j)!}\left(x\sqrt{2A}\right)^{m-2i}\left(x\sqrt{2B}\right)^{n-2j}. \tag{41}$$

Indeed, by (14), the integral representation (40) becomes

$$U_{m,n}(x,y,A,B)=\frac{1}{m!n!}\int_0^{\infty}e^{-t}t^{m+n}H_m\left(x,\frac{1}{t},A\right)H_n\left(y,\frac{1}{t},B\right)dt. \tag{42}$$

It is worthy to mention that, on taking n=0 or m=0, the Equations (40), (41) and (42) of the 2V2ICMP reduce to the Equations (38), (36) and (37) of the [CMPs], respectively.

It is evident that the formula (37) provides

$$U_n(x+z,A)=\frac{1}{n!}\int_0^{\infty}e^{-t}t^nH_n\left(x+z,\frac{1}{t},A\right)dt. \tag{43}$$

Thus, by applying (13) in (43), we obtain

$$U_n(x+z,A)=\sum_{k=0}^{n}\frac{1}{k!(n-k)!}\int_0^{\infty}e^{-t}t^nH_{n-k}\left(x,\frac{1}{2t},A\right)H_k\left(z,\frac{1}{2t},A\right)dt,$$

Which, in view of (14), one gets

$$U_n(x+z,A)=2^{-n/2}\sum_{k=0}^{n}\frac{1}{k!(n-k)!}\int_0^{\infty}e^{-t}t^{(n-k)+k}H_{n-k}\left(\sqrt{2}x,\frac{1}{t},A\right)H_k\left(\sqrt{2}z,\frac{1}{t},A\right)dt.$$

This, by the formula (42), leads to the addition theorem for the Chebyshev matrix polynomials of the second kind in the form

$$U_n(x+z,A)=2^{-n/2}\sum_{k=0}^{n}U_{n-k,k}\left(\sqrt{2}x,\sqrt{2}z,A,A\right). \tag{44}$$

GENERALIZED HERMITE MATRIX POLYNOMIALS

By using the hypergeometric matrix function it is convenient to consider a new generalized form of the Hermite matrix polynomials. The generalized Hermite matrix polynomials [GHMPs] of two matrices and two variables are presented here. Let A and B be two matrices in $C^{r\times r}$ such that A satisfies the condition (1) and B satisfies the condition (17). We can define the GHMPs in the form:

$$H_n^B(x,y,A) = n!\sum_{k=0}^{\lfloor n/2 \rfloor} \frac{(-1)^k}{k!(n-2k)!} B_{n,k}\left(x\sqrt{2A}\right)^{n-2k}, \tag{45}$$

Where

$$B_{n,k} = y^k F(-kI, B; -nI; y). \tag{46}$$

Note that, by (24), the expression (46) can be written in the form

$$B_{n,k} = \sum_{j=0}^{k} \frac{k!(n-j)!}{j!(k-j)!}(B)_j\, y^{k+j}. \tag{47}$$

when B is the zero matrix, then the GHMPs reduce to the two-variable Hermite matrix polynomials, $H_n^{\theta}(x,y,A) = H_n(x,y,A)$ .In view of (19), (20) and (34), the expression (46) can be also written in the following integral representation

$$B_{n,k} = y^k \Gamma^{-1}(B) \int_0^{\infty}\int_0^{\infty} e^{-(t+u)} t^n u^B \left(1 + \frac{yu}{t}\right)^k dt du. \tag{48}$$

It is clear that

$$H_n^B(-x,y,A) = (-1)^n H_n^B(x,y,A).$$

and

$$H_n^B(x,0,A) = \left(x\sqrt{2A}\right)^n.$$

By using (19), (20) and (34), the formula (45) leads to

$$H_n^B(x,y,A)=\Gamma^{-1}(B)\int_0^\infty\int_0^\infty t^n e^{-(t+u)}u^{B-I}\sum_{k=0}^{\lfloor n/2\rfloor}\frac{(-1)^k\left(x\sqrt{2A}\right)^{n-2k}}{k!(n-2k)!}y^k\left(1+\frac{yu}{t}\right)^k dtdu.$$

Hence, by (3), we obtain the integral representation of the GHMPs in the form

$$H_n^B(x,y,A)=\frac{\Gamma^{-1}(B)}{n!}\int_0^\infty\int_0^\infty t^n e^{-(t+u)}u^{B-I}H_n\left(x,y\left(1+\frac{yu}{t}\right),A\right)dtdu. \tag{49}$$

In view of (12), the integral representation (49) becomes

$$H_n^B(x,y,A)=\frac{\Gamma^{-1}(B)}{n!}\int_0^\infty e^{-u}u^{B-I}\int_0^\infty t^n e^{-t}H_n\left(x,\frac{y^2u}{t},A\right)dtdu\exp\left(-y(2A)^{-1}\frac{\partial^2}{\partial x^2}\right),$$

Which, by (37), provides the following form by means of the generalized Chebyshev matrix polynomials

$$H_n^B(x,y,A)\exp\left(y(2A)^{-1}\frac{\partial^2}{\partial x^2}\right)=\Gamma^{-1}(B)\int_0^\infty e^{-u}u^{B-I}U_n\left(x,y^2u,A\right)du. \tag{50}$$

Thus, by exploiting (19), (20) and (36) in (50), one gets

$$H_n^B(x,y,A)\exp\left(y(2A)^{-1}\frac{\partial^2}{\partial x^2}\right)=\sum_{k=0}^{\lfloor n/2\rfloor}\frac{(-1)^k(n-k)!y^{2k}}{k!(n-2k)!}(B)_k\left(x\sqrt{2A}\right)^{n-2k}.$$

By (11) and (6), it follows that

$$\begin{aligned}H_n^B(x,y,A)&=\sum_{k=0}^{\lfloor n/2\rfloor}(B)_k\frac{(n-k)!y^{2k}}{k!n!}\frac{(-1)^k n!}{(n-2k)!}H_{n-2k}(x,y,A)\\&=\sum_{k=0}^{\lfloor n/2\rfloor}(B)_k\left[(-nI)_k\right]^{-1}\frac{(-1)^k y^{2k}}{k!}\frac{\partial^k}{\partial y^k}H_n(x,y,A).\end{aligned}$$

Hence from (25) we arrive at the following representation of the GHMPs

$$H_n^B(x,y,A) = \Phi\left(B;-nI;-y^2\frac{\partial}{\partial y}\right)H_n(x,y,A). \tag{51}$$

The use of the second order matrix differential Equation (7) in the integral representation (49) gives

$$\frac{\Gamma^{-1}(B)}{n!}\int_0^\infty\int_0^\infty e^{-(t+u)}t^n u^{B-I}\left[y\left(1+\frac{yu}{t}\right)\frac{\partial^2}{\partial x^2}-xA\frac{\partial}{\partial x}+nA\right]H_n\left(x,y\left(1+\frac{yu}{t}\right),A\right)dt\,du=\theta,$$

Which, with the help of (5), obtaining the differential recurrence relation

$$\left[y\frac{\partial^2}{\partial x^2}-xA\frac{\partial}{\partial x}+nA\right]H_n^B(x,y,A)=B\frac{\partial}{\partial x}H_{n-1}^{B+I}(x,y,A)\sqrt{2A}.$$

REFERENCES

1. Jódar, L., Defez, E. and Ponsoda, E. (1996) Orthogonal Matrix Polynomials with Respect to Linear Matrix Moment Functionals: Theory and Applications. Approximation Theory and Its Applications, 12, 96-115.
2. Jódar, L. and Company, R. (1996) Hermite Matrix Polynomials and Second Order Matrix Differential Equations. Approximation Theory and Its Applications, 12, 20-30.
3. Jódar, L. and Defez, E. (1998) On Hermite Matrix Polynomials and Hermite Matrix Function. Approximation Theory and Its Applications, 14, 36-48.
4. Batahan, R.S. (2006) A New Extension of Hermite Matrix Polynomials and Its Applications. Linear Algebra and its Applications, 419, 82-92. http://dx.doi.org/10.1016/j.laa.2006.04.006
5. Batahan, R.S. and Bathanya, A.A. (2012) A New Generalization of Two-Variable Hermite Matrix Polynomials. Global Journal of Pure and Applied Mathematics, 8, 383-393.
6. Defez, E., Tung, M.M. and Sastre, J. (2011) Improvement on the Bound of Hermite Matrix Polynomials. Linear Algebra and its Applications, 434, 1910-1919. http://dx.doi.org/10.1016/j.laa.2010.12.015

7. Khan, S. and Raza, N. (2010) 2-Variable Generalized Hermite Matrix Polynomials and Lie Algebra Representation. Reports on Mathematical Physics, 66, 159-174. http://dx.doi.org/10.1016/j.laa.2010.12.015
8. Metwally, M.S. (2011) Operational Rules and Arbitrary Order Two-Index Two-Variable Hermite Matrix Generating Functions. Acta Mathematica Academiae Paedagogicae Nyegyhiensi (N.S.), 27, 41-49.
9. Metwally, M.S., Mohamed, M.T. and Shehata, A. (2009) Generalizations of Two-Index Two-Variable Hermite Matrix Polynomials. Demonstratio Mathematica, 42, 687-701.
10. Sayyed, K.A.M., Metwally, M.S. and Batahan, R.S. (2003) On Gegeralized Hermite Matrix Polynomials. Electronic Journal of Linear Algebra, 10, 272-279.
11. Shahwan, M.J.S. and Pathan, M.A. (2006) Origin of Certain Generating Relations of Hermite Matrix Functions from the View Point of Lie Algebra. Integral Transforms and Special Functions, 17, 734-747. http://dx.doi.org/10.1080/10652460600725069
12. Altin, A. and Çekim, B. (2012) Generating Matrix Functions for Chebyshev Matrix Polynomials of the Second Kind. Hacettepe Journal of Mathematics and Statistics, 41, 25-32.
13. Khan, S. and Al-Gonah, A.A. (2014) Multi-Variable Hermite Matrix Polynomials: Properties and Applications. Journal of Mathematical Analysis and Applications, 412, 222-235. http://dx.doi.org/10.1016/j.jmaa.2013.10.037
14. Dunford, N. and Schwartz, J. (1957) Linear Operators. 1, Interscience, New York.
15. Batahan, R.S. (2007) Generalized Gegenbauer Matrix Polynomials, Series Expansion and Some Properties. In: Ling, G.D., Ed., Linear Algebra Research Advances, Nova Science Publishers, 291-305.
16. Defez, E. and Jódar, L. (1998) Some Applications of the Hermite Matrix Polynomials Series Expansions. Journal of Computational and Applied Mathematics, 99, 105-117. http://dx.doi.org/10.1016/S0377-0427(98)00149-6
17. Hille, E. (1969) Lectures on Ordinary Differential Equations. Addison-Wesley, New York.
18. Jódar, L. and Cortés, J.C. (1998) Some Properties of Gamma and Beta Matrix Function. Applied Mathematics Letters, 11, 89-93. http://dx.doi.org/10.1016/S0893-9659(97)00139-0
19. Jódar, L. and Cortés, J.C. (1998) On the Hypergeometric Matrix Function. Journal of Computational and Applied Mathematics, 99, 205-217. http://dx.doi.org/10.1016/S0377-0427(98)00158-7
20. Defez, E. and Jódar, L. (2002) Chebyshev Matrix Polynomials and Second Order Matrix Differential Equations. Utilitas Mathematica, 61, 107-123.

CITATION

1. Batahan, R. (2014) Generalized Form of Hermite Matrix Polynomials via the Hypergeometric Matrix Function. Advances in Linear Algebra & Matrix Theory, 4, 134-141. doi: 10.4236/alamt.2014.42012.

Latent Structure Linear Regression

Agnar Höskuldsson
Centre for Advanced Data Analysis, Lyngby, Denmark

10

ABSTRACT

A short review is given of standard regression analysis. It is shown that the results presented by program packages are not always reliable. Here is presented a general framework for linear regression that includes most linear regression methods based on linear algebra. The H-principle of mathematical modelling is presented. It uses the analogy between the modelling task and measurement situation in quantum mechanics. The principle states that the modelling task should be carried out in steps where at each step an optimal balance should be determined between the value of the objective function, the fit, and the associated precision. H-methods are different methods to carry out the modelling task based on recommendations of the H-principle. They have been applied to different types of data. In general, they provide better predictions than linear regression methods in the literature.

INTRODUCTION

Regression methods are among the most studied methods within theoretical statistics. Numerous books have been published on the topic. They are standard in mathematical classes in most universities in the world. Advanced program packages like SAS and SPSS have been used by students since the 1970s. There are still great interest and developments in regression methods. The reason is that many modern measurement equipments require special methods in order to carry out regression analysis.

In applied sciences and industry there are three fundamental issues.

- Data have latent structure. Data in applied sciences and industry have typically latent structure. This means that values of the variables are geometrically located in a low-dimensional space. The location is often a low dimensional ellipsoid. Variables are often correlated and therefore, may or should not be treated as independent. Mathematical models and methods that assume that X-data are of full rank are often incorrect.
- It is a tradition to compute optimal or unbiased solutions. The optimal solution has often interesting properties, when data satisfy the given model. But in practice the optimal solution has often bad or no prediction ability. It means that the model cannot be used for predictions at new samples. By relaxing on the optimality or unbiasedness a better solution can often be obtained.
- Statistical significance testing. Sometimes the significance testing in linear regression analysis may not be reliable. This is a serious problem that is being discussed at professional organizations. See Section 2.3.

In Section 2, a brief review is given of linear least squares methods for standard linear regression. The basic properties of the modelling task are discussed. In Section 3, a general framework for standard linear regression is presented. The basic idea is to separate how one should view the regression task and the associated numerical procedure. The numerical part is the same for all types of linear regressions within this framework. It allows flexibility in developing new methods that fit different types of data and objectives. In Section 4, the H-principle is formulated. It uses the analogy between the modeling task and measurement situation in quantum mechanics. It assumes that data are centered and scaled. It suggests that one should use the covariances as the weight vectors in the general algorithm. Section 5 considers some methods to validate latent structure results. In Section 6, a brief introduction is given to H-methods. They focus on finding what part of data should be used, when covariances are used as weight vectors. In Section 7, the use of an H-method is compared to Ridge Regression. The results obtained look similar, but the H-method uses much smaller dimension. In Section 8, the use of latent structure linear regression is discussed. It is shown that the latent structure methods much more quickly arrive at the final solution than forward variable selection.

Notation. matrices are denoted by upper-case bold letters and vectors by lower-case bold letters. The letters i, j and k denote indices relative to vectors and matrices, while a and b are related to the steps in the methods. The regression data are (X, y), X is N × K matrix and y an N-vector. X is the instrumental data and y the response data. Only one y-variable is considered here in order to simplify the notation. Data are assumed to be centred, i.e. mean values subtracted from columns of X and y.

LINEAR LEAST SQUARES METHODS

The Linear Model and Solution

It is assumed that there are given instrumental data X and response data y. A linear regression model is given by

$$\boldsymbol{y} = \beta_1 \boldsymbol{x}_1 + \beta_2 \boldsymbol{x}_2 + \cdots + \beta_K \boldsymbol{x}_K + \boldsymbol{\varepsilon} \tag{1}$$

The x-variables are called independent variables and the y-variable the dependent one. When the parameters β have been estimated as b, the estimated model is now

$$y = b_1 x_1 + b_2 x_2 + \cdots + b_K x_K \tag{2}$$

when there is given a new sample $x_0 = (x_{10}, x_{20}, L, x_{K0})$, it gives the estimated or predicted value $y_0 = b_1 x_{10} + L + b_K x_{K0}$. Greek letters are used for the theoretical parameters and roman letters for the estimated values, because there can be many estimates for β. Linear least squares method is usually used for estimating the parameters. It is based on minimizing the residuals, $(y\text{-}Xb)^T(y\text{-}Xb) \rightarrow \text{minimum}$. In case the data follow a normal distribution, the procedure coincides with the Maximum Likelihood method. The solution is given by

$$\boldsymbol{b} = \left(\boldsymbol{X}^T\boldsymbol{X}\right)^{-1}\boldsymbol{X}^T\boldsymbol{y} \tag{3}$$

The assumption of normality is often written as $y \sim N\left(X\beta, \sigma^2\right)$. This means that the expected values of y is E(y) = Xβ and the variance is $\text{Var}(y) = \sigma^2 I$, I is the identity matrix. The parameters b have the variance given by,

$$\text{Var}(\boldsymbol{b}) = \sigma^2\left(\boldsymbol{X}^T\boldsymbol{X}\right)^{-1} \tag{4}$$

Here σ^2 is estimated by

$$\sigma^2 \cong \sum_1^N e_i^2 / (N - K) \tag{5}$$

where the residuals, e_i's, are computed from e = y – Xb. The estimates (3) have some desirable properties compared with other possible estimates:

- The estimate b is unbiased, E(b) = β.
- The estimate b has the minimum variance among all linear estimates of β, which are unbiased. It means that if another unbiased estimate of β has a variance matrix E, then E–Var(b) is a semi-positive definite matrix.

The matrix $(X^TX)^{-1}$ is called the precision matrix, and it shows how precise the estimates bare. We would like the variance matrix Var(b) to be as small as possible. a) and b) state that assuming the linear model (1), the solution (3) is in fact theoretically the best possible one. It happens often that the model (1) is too detailed in the sense that it contains many variables. In this case a significance test can be carried out to evaluate if some variables in the model (1) are not significant, and thus may be excluded from the model. Interpretation of the estimation results is an important part of the statistical analysis. A parameter value is evaluated by a t-test. A t-test of a parameter b_i is given by

$$t = \frac{b_i}{\sqrt{\mathrm{Var}(b_i)}} \cong \frac{b_i}{\sqrt{\left(s^2 \times s^{ii}\right)}} \tag{6}$$

Here s^2 is given by (5) and s^{ii} is the ith diagonal element of $(X^TX)^{-1}$.

The Linear Least Squares Solution in Practice

The model (1) is so obviously proper that statisticians do not doubt its validity. In the classroom it is emphasised that an important part of the statistical analysis is the interpretation of the resulting equation (2). Changes in y can be related to changes in x_i, for instance, $\Delta y = b_i \Delta x_i$.

But the model (1) is incorrect in most cases in applied sciences and industry (apart from simple laboratory experiments) and the nice theory above is not applicable. Why is this the case? We cannot automatically insert new samples into (2) and compute the y-value. It must be secured that the sample used is located along the samples used for analysis, the rows of X. A correct model is

$$y = \alpha_1 \tau_1 + \alpha_2 \tau_2 + \cdots + \alpha_A \tau_A + \varepsilon. \tag{7}$$

Here (t_a) are latent variables and (a_a) the regression coefficients on the latent variables. Latent variables are linear combinations of the original variables. Thus the regression task is both to find the latent variables, τ's, and the associated regression coefficients. The resulting equation (7) is then converted into (2). The model (7) is called latent structure linear regression. The data values associated with a latent variable are called score values.

The problem in using the least squares solution appears, when the precision matrix becomes close to singular. Program packages like SAS and SPSS inform the user when it has become close to being singular.

But the information is concerned with determining the precision matrix by the numerical precision of the computer. There are practical problems long before the algorithm informs that the precision matrix is close to be singular and estimates, b's, may not be determined precisely.

Many statisticians do not like latent structure regression. Singularity is also well understood among statisticians. When singularity occurs many statisticians recommend Ridge Regression, Principal Component Regression (PCR), variable selection methods or some other "well understood" methods.

As mentioned above, statisticians emphasize the interpretation of parameters of the results (2). But this is difficult, if not impossible, in most latent structure regressions. The measurement data often represent a "system", where variables are dependent on each other from chemical equilibrium, action-reaction in physics and physical and technical balance. In applied sciences and industry the emphasis is on prediction. "The quality of models is to be judged by their predictions." A good example is the company Foss Analytic, Denmark, which produces measurement instruments for the food and oil industry and has sales of around 200 mio euros pr year. The instruments are calibrated by regression methods and the company is only interested in how well the instruments measure new samples.

Significance Testing

Program packages, SAS, SPSS and others, introduce the results of regression analysis by presenting the regression coefficient, b_i, its standard deviation, a t-test value for its significance, (6), and the significance probability. These results are typically copied into the scientific report and added *'s to indicate how significant the parameters are (One * is 5% significance, etc.). This report is then basis for further publications. But this analysis may not be reliable. Consider the t-test closer and take the last variable, x_K. Suppose that we for the data can write

$$x_K = z_K + t_K, \tag{8}$$

where t_K is orthogonal to the other variables data, x_1, x_2, L, x_{K-1}. This can be achieved by Gram-Schmidt orthogonalisation of X. Then the t-value (6) for the parameter b_K can be written as

$$t = \frac{y^T t_K}{\left(s \times |t_K|\right)} \tag{9}$$

This shows that the t-value is invariant to the size of t_K, which is the marginal value of x_K given the other variables. The size of t_K can be small and without any practical importance, while being significant by (9). When there are many variables, there will be a few, where the marginal effect is small but statistically significant. Therefore, significance testing in the experimental studies will usually declare a few variables significant, although they have no practical importance. This has made the American Psychological Association introduce the rule that papers that only base their results on significance testing are no longer accepted in scientific psychological journals. Instead guidelines for experimenters have been prepared.

Some Aspects of the Modelling Task

It is instructive to study closer the variance associated with new samples in the linear least squares model. Suppose that x_0 is a new sample and the associated response value y_0 is computed by (2). Using (4) and (5) the variance of y_0 can be written as

$$\mathrm{Var}(y_0) \cong \left[y^T y - y^T X\left(X^T X\right)^{-1} X^T y\right] \times \left[1 + x_0^T \left(X^T X\right)^{-1} x_0\right] \Big/ (N - K) \tag{10}$$

It can be stated that the primary objective of the modelling task is to keep the value of (10) as small as possible. There are two aspects of (10) that are important. When a new variable (or component) is added to the model, it can be shown that 1) the error of fit, $\left[y^T y - y^T X\left(X^T X\right)^{-1} X^T y\right]$, always decreases 2) the model variation, $\left[1 + x_0^T \left(X^T X\right)^{-1} x_0\right]$, always increases

(In theory these measures can be unchanged, but in practice these changes always occur). If the data follow a normal distribution, it can be shown that a): the error of fit, $\left[y^Ty - y^TX\left(X^TX\right)^{-1}X^Ty\right]$ b): the precision matrix, $\left(X^TX\right)^{-1}$ are stochastically independent. It means that the knowledge of (a) does not give any information on (b). In order to know the quality of predictions, (b) must be computed. Program packages use the t-test or some equivalent measure to test the dimension. No information is given on the precision matrix. It must be computed separately in order to find out how well the model is performing. But a modelling procedure must include both (a) and (b) in order to secure small value of (10). It must have a procedure of how to handle the decrease of 1) and the increase of 2) during the model estimation.

FRAMEWORK FOR LINEAR REGRESSION ANALYSIS

Views on the Regression Analysis

Here is presented a general framework for carrying out linear regression. The basic idea is to separate the computations into two parts. One part is concerned how it is preferable to look at data and the other part the numerical algorithm to compute the solution vector and associated measures. At the first part it is assumed that a weight matrix $W_K = (w_1, L, w_K)$ is given. Each weight vector should be of length one, although this is not necessary. The weight vectors (w_a) reflect how one wants to look at the data. If $w_a = (0, L, 0, 1, 0, L, 0)$, a variable is selected according to some criterion. For w_a equal the ath eigenvector of X^TX we get PCA. If w_a is the left singular vector of $X^T_{a-1}Y_{a-1}$, the result will be PLS regression. Some program packages have the option of defining priorities for the variables, where higher priority variables are treated first. This can be achieved by defining elements of w_a to be zeros for variables that are not used at this step. Different options can be mixed in order to define the weight vectors that prescribe how the regression analysis should be carried out. The role of the weight vectors is only to compute X-score vectors, when instrumental data are given and to compute loading vectors when covariances are given. The only requirement to the weight vectors is that the resulting score, or loading vectors, may not be zero. The other part of the computations is a numerical algorithm, which is the same for all admissible choices of W_K.

A Numeric Algorithm

In most cases in regression it is assumed that there are given instrumental data X and response values y. In Ridge regression and engineering applications like Kalman filtering and some others, the starting point is a covariance matrix S. In Ridge Regression $S=X^TX+kI$, where k is a constant determined by data.

Here the start of the algorithm can be of two kinds, which are treated separately.

Instrumental data

The equations are written for several y-vectors. Initially $X_0 = X$, $Y_0 = Y$, $S_0 = S = X^TX$, $C_0 = C = X^TY$, $S^+ = 0$ and $B = 0$. For a=1, L, K:

Score vector :$t_a=X_{a-1}w_a$

Loading vector:$p_a=X_{a-1}^T t_a$

Scaling constant:$d_a= 1/|t_a|^2$

Y-loading vector:$q_a=Y_{a-1}^T t_a$

Loading weight vector:v_a

The loading weight vectors (v_a) are defined by thssse equation T = XV, or $t_a = Xv_a$. X and Y are adjusted at each step by the score vector t_a,

$$\boldsymbol{X}_a = \boldsymbol{X}_{a-1} - d_a \boldsymbol{t}_a \boldsymbol{p}_a^T, \quad \boldsymbol{Y}_a = \boldsymbol{Y}_{a-1} - d_a \boldsymbol{t}_a \boldsymbol{q}_a^T$$

Variance and covariances are given

Initially $S_0 = S$ and $C_0 = C$.

Initially S_0=S and C_0=C.

Loading vector: $p_a = S_{a-1} w_a$

Scaling constant: $d_a = 1/(w_a^T p_a)$

Y-loading vector: $q_a = C_{a-1}^T w_a$

Loading weight vector : v_a

The loading weight vectors are defined by the equation P = SV, or $p_a = Sv_a$. S is adjusted by the loading vector p_a

$$S_a = S_{a-1} - d_a p_a p_a^T$$

Continuation of the Algorithm

In either case the continuation is as follows. The loading weight vectors are computed by

$$v_1 = w_1, \quad v_a = w_a - d_1 \left(p_1^T w_a\right) v_1 - \cdots - d_{a-1} \left(p_{a-1}^T w_a\right) v_{a-1} \quad a = 2,3,\cdots$$

The covariance matrix C is adjusted and regression coefficients and precision matrix estimated,

$$C_a = C_{a-1} - d_a p_a q_a^T, \quad B_a = B_{a-1} + d_a v_a q_a^T, \quad S_a^+ = S_{a-1}^+ - d_a v_a v_a^T$$

The expansions that are being carried out by the algorithm are:

$$X = d_1 t_1 p_1^T + \cdots + d_A t_A p_A^T + \cdots + d_K t_K p_K^T = TDP^T$$

$$\hat{Y} = d_1 t_1 q_1^T + \cdots + d_A t_A q_A^T + \cdots + d_K t_K q_K^T = TDQ^T$$

$$X^T Y = d_1 p_1 q_1^T + \cdots + d_A p_A q_A^T + \cdots + d_K p_K q_K^T = PDQ^T$$

$$\hat{Y}^{T}\hat{Y} = d_1 q_1 q_1^{T} + \cdots + d_A q_A q_A^{T} + \cdots + d_K q_K q_K^{T} = QDQ^{T}$$

$$S = d_1 p_1 p_1^{T} + \cdots + d_A p_A p_A^{T} + \cdots + d_K p_K p_K^{T} = PDP^{T},$$

$$B = S^{-1}X^{T}Y = d_1 v_1 q_1^{T} + \cdots + d_A v_A q_A^{T} + \cdots + d_K v_K q_K^{T} = VDQ^{T},$$

$$S^{-1} = d_1 v_1 v_1^{T} + \cdots + d_A v_A v_A^{T} + \cdots + d_K v_K v_K^{T} = VDV^{T}$$

Normally, only A terms of the expansions are used because it is verified that the modelling task cannot be improved beyond A terms.

The numerical procedure has the properties:

The adjustments of X or S are rank one reductions. In the case of instrumental data the score vectors (t_a) are orthogonal. (v_a) and (p_a) are conjugate sets, i.e., $v_a^{T}p_b = 0$ for $a \neq b$, and $v_a^{T}p_a = 1/d_a$. In matrix term $V^{T}P = D^{-1}$. For proof of these statements and some more properties, which can be stated in general for admissible weight vectors (w_a), see [1].

Numerical precision is important, when writing general computing programs. The adjustments can be numerically unstable. Also, inner products can be unstable if the number of multiplications is large. Therefore, the score vector t should be scaled to unit length before being used.

If one is working with a test set, X_t, the estimated y-values can be computed as $\hat{Y}_t = X_t B$. The score vectors for the test set are computed as $T_t = X_t V$. We can plot columns of $\hat{Y}_t$ against the score vectors of the test set, the columns of T_t.

Note, that the weight vectors (w_a) do not appear in the expansions. They are only used to determine the score or loading vectors. The algorithm is independent of the size of the weight vectors. But in order to secure numerical precision it is recommended to scale the vectors to be of unit length. The number of terms, A, in the expansions will depend on the set of w_a's being used.

THE H-PRINCIPLE OF MATHEMATICAL MODELLING

In the 1920s W. Heisenberg pointed out that in quantum mechanics there are certain magnitudes that are conjugate and cannot be determined exactly at the same time. He formulated his famous uncertainty inequality, which states that there is a lower limit to how well the conjugate magnitudes can be determined at the same time. The lower limit is related to the Planck's constant of light. An example of such conjugate magnitudes is the position and momentum (speed) of an elementary particle. The inequality is formulated as

$$\Delta(\text{position})\times\Delta(\text{momentum}) \geq \text{constant}$$

N. Bohr pointed out that there are many complementary magnitudes, which cannot be determined exactly at the same time. In practice it means that the experimenter must be aware of that the instrument and the phenomenon in question place some restrictions on the outcome of the experiment. It is only by the application of the instrument that the restrictions are detected. The uncertainty inequality is a kind of guidance of what can be expected.

The H-Principle

When modelling data we have analogous situation. Instead of a measurement instrument we have a mathematical method. ΔFit and ΔPrecision are conjugate magnitudes that cannot be controlled at the same time. It is necessary to carry out the modelling in steps and at each step evaluate the situation as prescribed by the uncertainty inequality. The recommendations of the H-principle are:

1) Carry out the modelling in steps. You specify how you want to look at the data at this step by formulating how the weights are computed.
2) At each step compute
   - expression for improvement in fit, Δ(Fit)
   - and the associated prediction, Δ(Precision)
3) Compute the solution that minimizes the product Δ(Fit)×Δ(Precision)

4) In case the computed solution improves the prediction abilities of the model, the solution is accepted. If the solution does not provide this improvement, the modelling is stopped.
5) The data is adjusted for what has been selected; restart at 1.

Consider closer how this applies to linear regression. The task is to determine a weight vector w according to this principle. For the score vector t = Xw we have

- Improvement in fit: $|Y^T t|^2 / (t^T t)$
- Associated variance: $\Sigma / (t^T t)$

Treating S as a constant the task is to maximize

$$\left[\frac{|Y^T t|^2}{t^T t}\right] \times \frac{1}{\left[\frac{1}{t^T t}\right]} = w^T X^T Y Y^T X w$$

This is a maximization task because improvement in fit is negative. The maximization is carried out under the restriction that w is of length 1, |w| = 1. Using Lagrange multiplier method it can be shown that the task is an eigenvalue task,

$$X^T Y Y^T X w = \lambda w$$

In case there is only one y-variable, the eigenvalue task has a direct solution

$$w = X^T y / |X^T y|$$

These are the solutions used in PLS regression. Thus we can state that PLS regression is consistent with the H-principle. Thus, the H-principle in case of general linear regression suggests:

- Carry out the modelling task in steps.
- It recommends using as weight vectors in the algorithm in Section 3 the covariances of the reduced data,

$$w_a \propto X_{a-1}^{T} y_{a-1} \tag{11}$$

In case of instrumental data (X, y) it is recommended to use (11), which will give standard PLS regression. In case some variables are not to be included, the indices in w_a associated with these variables are zero. In case that variance matrix S is used as a starting point, like at Ridge Regression, the solution is found by using (11) as weight vectors.

Interpretation in Terms of Prediction Variance

Suppose that a score vector t = Xw has been computed. The explained fit by the score vector is $(y^T t)^2/(t^T t)$. If x_0 is a new sample, the variance of the estimated y-value would be

$$\mathrm{Var}(y(x_0)) \cong \left[y^T y - \frac{(y^T t)^2}{t^T t} \right] \times \left[1 + \frac{t_0^2}{t^T t} \right] \Big/ (N-1)$$

Here t_0 is the score value associated with x_0, $t_0 = x_0^T w$. The properties of the normal distribution are being used here. The latent variable associated with t will be normally distributed, because it is a linear combination of the original variables. The expression for the variance is a conditional variance given the score vector. Define now the function

$$f(w) = \left[y^T y - \frac{(y^T t)^2}{t^T t} \right] \times \left[1 + \frac{t_0^2}{t^T t} \right]$$

The function f has the property that it does not change when w is scaled by a positive constant, c > 0. Choose $c = c_0$ so that $|t| = 1$. Then we can write

$$f(w) = f(c_0 w) = \left[y^T y - \left(y^T t \right)^2 \right] \times \left[1 + t_0^2 \right]$$

Treating t_0 as a constant it is clear that maximizing the covariance is important.

Note that the argumentation used here is heuristic, because we use the scale invariance of f(w) to assume |t| = 1.

Scaling of Data

The H-principle and associated H-methods assume that the columns of X and Y have equal weight in the analysis. Therefore, it may be necessary to scale the data. Scaling of columns of X and Y can be achieved by multiplying the matrices from the right by a diagonal matrix. The linear least squares solution is given by

$B = \left(X^T X \right)^{-1} X^T Y$ .If X and Y are scaled column-wise (by variables), it amounts to the transformations $X \leftarrow (XC_1)$ and $Y \leftarrow (YC_2)$, where C_1 and C_2 are diagonal matrices. The solution B can be obtained from the solution for scaled data, B_1, as follows,

$$B = C_1 \left[\left\{ (XC_1)^T (XC_1) \right\}^{-1} (XC_1)^T (YC_2) \right] C_2^{-1} = C_1 B_1 C_2^{-1}$$

This equation shows that if we are computing or approximating the linear least squares solution, we can work with scaled data. When we want the solution for the original data, we scale "back" as shown in the equation. This property is also used, when the approximate solution is being computed.

The effect of scaling is a better numerical precision. In the case of spectral data scaling is usually not necessary. If it is desired to work with the differential or the second differential of the spectral curve, it may be necessary to scale the data in order to obtain better precision.

If original data follow a normal distribution, the scaled ones will not. There will be a slight change due to scaling. But properties due to normality are only used as here as guidance. Slight deviations from normality will not affect the analysis.

Scaling of data is a difficult subject. e.g., if some variables have many values below the noise level, it may be risky to scale data. The problem of scaling is not restricted to specific methods of regression analysis. This topic is not considered closer here.

H-Principle in Other Areas

The H-principle has been extended to many areas of applied mathematics. In many cases it has opened up for new mathematics. In [2] it has been extended to path modelling, giving new methods to carry out modelling of data paths. In [3] it has been applied to non-linear modelling that may give good low rank solutions, where full rank regularized solutions do not give convergence. In [4] it is extended to multi-linear algebra, where there are many indices in data, e.g., $X = (x_{ijklm})$ and $y = (y_{ij})$. The use of directional inverse makes it possible to extend methods of ordinary matrix analysis. It has also been applied to dynamic systems (process control and time series), pattern recognition and discriminant analysis, optimisation (quadratic and linear programming) and several other areas. By working with latent variables they can be "tailored" to handle the mathematical objective in question. e.g., dynamic systems are typically estimated by linear least squares methods. But the H-principle suggests determining score vectors with optimal forecasting properties resulting in better methods than traditional ones. Numerical analysts talk about an ill-posed or ill-conditioned mathematical model, when the precision matrix is singular or close to singular. But there can be a stable low rank solution, which is fine to use. This is especially important, where interand extrapolation is needed for the solution obtained (e.g. at partial differential equation).

MODEL VALIDATION

Many methods in statistical analysis are based on search or some optimisation procedure. But the results are evaluated as if they are based on random sampling. For instance, in (forward) stepwise regression a search is carried out to find the variable that has the largest squared correlation coefficient with the response variable. The significance of the finding is then evaluated by a t-test as if the variable was given beforehand. But in random data we can expect 5% of the variable to be significant by a statistical test. In PLS regression a weight vector is determined by maximizing the size of the resulting y-loading vector, q. In these cases it is important to validate the obtained results. There are many criteria available to check the dimension of a model. Examples are Mallows' C_p value and Akaike's information criterion. But they are developed for random sampling from a distribution. In practice they suggest too high number of (latent) variables. Therefore, especially in chemometrics, the practice has become to use cross-validation and a test set. Experience suggests that both should be used. It is recommended at each step to look at how well the score vectors are doing. This holds for both calibration data, which are used for estimating the parameters, for the test set and for cross-validation. Thus, we plot y against the columns of T, the test set y_t against T_t and cross-validated y-values, y_c, against T_c. If the last two types of plots do not show any correlation, it is time to consider stopping modelling of data.

In most cases these procedures work well. But there may be cases when they do not work well. This may happen, when there is nonlinearity in data, some extreme y-values, gaps in data or grouping in data. Cross validation or the use of a test set may be affected by these special features in data. It can be recommended to check the Y-t (y against the columns of T) plots to see if there are these special features in data. Simple statistical tests can also be used to inform of these special features. An important test is a test of randomness of residuals. But many other can be recommended.

Cross-Validation

At cross-validation the samples are divided into groups, say 10 groups. One group is put aside and the other 9 are used to estimate the parameters of the model. This is repeated for all groups. The result is an estimation of the y-values, $\hat{Y}_c$, where each is estimated by using 90% of the samples. Similarly, score values, T_c, are estimated at the same time as the y-values similarly as at a test set, see Section 3.2. In statistics it is common to use leave-one-out cross-validation, where the number of groups is equal to the number of samples. This procedure is important to use in order to check for outliers, which may spoil the modelling task. But it is not considered very efficient as a validation procedure. In chemometrics it is common to use 7 fold cross-validation where samples are randomly divided into 7 groups. The argument is that the model developed needs a strong evaluation test.

There may be problems, when using the cross-validation approach. In the glucose data considered here, around 80% of the glucose values are in the range 3.5 –7. The histogram of the larger values falls smoothly until 23.5. Even the histogram of the logarithmic values is skew. Different cross-validations give different results depending on how the large y-values are located in the groups. Therefore, it may be better to use ordered cross-validation. Here the y-values are ordered and say, every 10th of the ordered y-values define the groups. In the examples in Section 7 this type of ordered cross-validation is used (first group 1st, 11th, ..., etc).

Test Set

Industry standards [5] require that a new measurement method should be tested by 40 new samples in a blind test. In a blind test the 40 samples have not been used in any way in the modelling task. Ideally, sufficient amount of samples should be obtained so that 40 samples can be put aside. Test set is often selected by ordering the y-values like at ordered cross-validation. However, it may be better to use the Kennard-Stone procedure, [6] , that orders the samples according to their mutual distance. The first two are the ones that are most apart, next sample the furthermost away from the first two, etc. In the examples in this paper a test set is determined by selecting every 5th of the ordered samples by the Kennard-Stone procedure.

Sometimes, different types of test sets should be selected. As an example one can mention the situation, where most y-values are very small and only few are large. Here it may be better to order the samples according to the first PCA or PLS score vectors.

Measures of Importance

Modelling of data should continue as long as there is covariance. The dimension of a model is determined by finding, when $X^T y \cong 0$ for reduced matrices. There are two ways to look at this. One is to study the individual terms, $(x_i^T y)$, and the other is total value, $y^T X X y^T = \Sigma_i (x_i^T y)^2$. Assume that data can be described by a multivariate normal distribution with a covariance matrix S_{xy}. Then, it is shown in [7] that the sample covariances $(x_i^T y)/(N-3)$ are approximately normally distributed. If $s_{xi,y} = 0$, it is shown in [7] that approximate 95% limits for the residual covariance, $(x_i^T y)/(N-3)$, are given by

$$\pm 1.96\sqrt{N}\sigma_{xi}\sigma_y/(N-3) \cong \pm 1.96\sqrt{N}s_{xi}s_y/(N-3)$$

Thus, when modelling stops, it is required that all residual covariances should be within these limits. If $s_{xi,y} = 0$, the distribution of the residual covariance approaches quickly the normal distribution by the central limit theorem. Therefore, it is a reliable measure to judge, if the covariances have become zero or close to zero.

If the covariance S_{xy} is zero, $S_{xy} = 0$, the mean, $E\{(y^T X X^T y)\}$, and variance, $Var\{(y^T X X^T y)\}$, can be computed, see [1]. The upper 95% limit of a normal distribution $N(\mu, s^2)$ is $\mu + 1.65\sigma$. This is used for mean and variance. In the analysis it is checked if $y^T X X^T y$ is below the upper 95% limit (a one-sided test),

$$y^T XX^T y < \mathrm{tr}\left(X^T X\right)\left(y^T y\right)/N + 1.65\sqrt{2\mathrm{tr}\left(X^T XX^T X\right)\left(y^T y\right)^2 / N^2}$$

tr is the trace function. When this inequality is satisfied, there is an indication that modelling should stop. This analysis has been found useful, when analysing spectral data.

H-METHODS

The H-principle recommends using as weights in the regression algorithm the covariances at each step, $w_a \propto X_{a-1}^T y_{a-1}$. This holds both in the case where there are given instrumental data (X,y) and variance/covariance (S, $X^T y$). H-methods study the situation, when the weight vectors are the (residual) covariances. They are of two types. One type is concerning finding the data, $X_1 = (x_{i1}, x_{i2}, \ldots; x_{in})$, that should be used. The other type are studies of the weights, (w_a). A brief review of some of the methods is presented.

Determining a Subset of Variables

It is important to find a good subset of variables for use in regression analysis. Some of the most used methods in program packages are variable selection methods. In a forward selection method variables are selected one by one. The variable that is added to the pool of selected variables is the one the gives the highest value of some measure among the variables that have not been selected.

Forward and Backward Selection of Variables

In latent structure regression the following approach has been found efficient. Carry out a forward selection of variables. A variable is added to the pool of selected variables, if it, together with the variables already selected, gives the best cross-validation among the unselected variables. Backward selection of variables to be deleted is then started, where the best results of the forward selection were obtained. One variable is deleted from the pool of variables. The variable is deleted, which gives the highest cross-validation values among the variables in the pool. The number of variables chosen for analysis is the one that gives the highest value of cross-validation at the backwards deletion. This procedure can be time consuming. Altogether K(K + 1)/2 forward regression analysis with cross-validation are carried out. If K = 1000 more than half a million analysis is carried out.

Analysis Based on Principal Variables

Principal variables are ones that have maximal covariance. The first one is the one at max $\left|(x_i^T y)\right|$. In the algorithm $w = (0, 0, ..., 1, 0, ...)$, where the index of 1 is at the principal variable, [1]. The next principal variable is the one having maximal covariance, now for the adjusted data. The important property is that if the covariance is not small, the size of the reduced x_i is not small, $\left|(x_i^T y)\right| \leq |x_i||y|$. Principal variables are used to replace the original ones with a smaller set of variables. e.g. a spectrum may represent 1000 variables. Computing the principal variables, one may suffice to analyse, say 150 principal variables instead of the original 1000. Forward and backward selection of variables can be carried out now for the principal variables. It is known that it is not efficient to base the modelling task on the first few principal variables, although this is stated in [8].

Modelling along Ordering of Variables

There are now available spectroscopic sensors that can be installed at different locations in a company. Especially chemical process companies have installed many; 200 locations is not large. There is need for a simple method that can be programmed easily and is not time consuming to run. This can be obtained by computing the simple correlations coefficients, r_{yxi}, or the covariances, $x_i^T y$, and sort them according to their numerical value with largest first. PLS regression can be carried out along this ordering of variables. This simple method, which is described in [9], has been implemented at several chemical process companies.

Study of the Residual Covariances

Small values in weight vectors indicate that corresponding variables have small influence on the results. For spectroscopic data there always are a collection of variables that should not be used. There can also be a question of the focus of the regression analysis that may suggest some other weight values then (11).

Cov-Proc Methods

In bio-assay studies there can be 3000 –4000 variables or even more. The experimenter is equally interested in important variables and as in prediction. This is done by the CovProc method, [10]. The idea is to combine the approach of theoretical statistics, where focus is on R^2, and of PLS regression, where score vector may not be too small. When the weight vector, w, of PLS regression has been computed, it is sorted according to size, $w^{(s)}$, the largest first. A score vector is chosen that gives the maximum of R^2,

$$\boldsymbol{t}_i = \boldsymbol{X}^{(s)}\left(w_1^{(s)}, w_2^{(s)}, \cdots, w_i^{(s)}, 0, 0, \cdots, 0\right),$$

$$\max R_i^2 = \max\left(\boldsymbol{y}^{\mathrm{T}}\boldsymbol{t}_i\right)^2 \Big/ \left(\boldsymbol{t}_i^{\mathrm{T}}\boldsymbol{t}_i\right), i = 1, 2, \cdots, K$$

$X^{(s)}$ is the same sorting of columns of X. CovProc method has the advantage that the first few score vectors, typically one or two or three, are based on very few variables but they explain large variation of X and Y. This implies that plots of the y-t_1, y-t_2, y-t_3 and t_1-t_2-t_3 can be transferred to few variables, which makes the interpretation useful. Large experimental studies have found this method efficient, [11]. Different comparisons to other methods in the experimental area have been carried out, see [12] [13], where CovProc method is the preferred one.

Small Values of Residual Covariances

When there are many variables, it is necessary to delete variables that have no or small covariance with the response variable. A useful procedure is to sort the values of the weight vectors, w_a, and study the small values found. One procedure is to delete one by one of the smallest values and measure the effects. This is continued until no more should be deleted. Then we start over with the deleted values and investigate if some of the deleted ones should be included again. This procedure is a "fine tuning" of the weights. If no other procedure is used, it often gives significant improvement compared to the original weights.

COMPARISON OF METHODS

The Data

210 blood plasma samples were measured on Bruker Tensor 27 FTIR instrument, [14]. The measurement unit that receives the plasma samples is AquaSpec from Micro-Biolytics, [15]. The direct measurements of glucose in blood plasma samples was carried out on a Cobas C 501 instrument by staff members at Department of Clinical Biochemistry, Holbæk Hospital, Denmark. Most of the samples were from diabetes patients giving high glucose values. The producer of the Cobas instrument that measured glucose, Roche Diagnostics D-68298 Mannheim, informs that the coefficient of variation, CV, is less than 1.5%. The average glucose value is 8.17 mmol/l. This gives a standard deviation of s = 0.12. Thus, we may expect that the laboratory values of glucose have uncertainties with this standard deviation.

Each spectrum from the Bruker instrument consists of 1115 values. The differential spectra were used. Cross-validation showed that better results were obtained by using the differential spectra compared to the original ones. The instrumental data X are thus 210 × 1115 and the y-data a vector of 210 glucose values. Leave-one-out analysis revealed that one sample was an outlier and it was removed. Forward selection of variables was carried out followed by backwards deletion of variables, Section 6.1.1. A measure of cross-validation was used in selection and deletion of variables. The backward analysis started at the number of variables, where the cross-validation at the forward analysis was at its highest. The result was that 50 variables of the original 1115 were optimal to use. Thus, in the following analysis X is 209 × 50 and y is a vector of 209 elements. In each analysis, both forward and backward, PLS Regressions are used with dimension A at most 20.

PLS Regression

The data were centred and scaled to unit variance. Considerable improvement was obtained by using scaling of spectra. The methods of section 5.3 suggested that the covariances had become zeros, when the 12th dimension had been completed. Analysis of cross-validations and test sets also suggested dimension 12.

In Figure 1(a) is shown a scatter plot between the measured glucose values on y-axis and the estimated y-values $\hat{Y}$ on the x-axis by PLS regression having dimension 12. The squared correlation coefficient of the scatter of points is $R^2 = 0.9980$ and standard deviation of the residual deviation, $(y-\hat{Y})$, equals s = 0.170 mmol/L. In Figure 1(b) is shown the scatter plot of glucose values on y-axis and the cross-validated y-values, $\hat{Y}_c$, on x-axis. Ordered cross-validation using the y-values was used, see Section 5.1. Here the squared correlation of the scatter of points is $R^2 = 0.9970$ and the standard deviation of $(y - \hat{Y}_c)$ equals s = 0.208 mmol/L. Figures 1(c) and (d) are the results of applying a test set. 40 samples derived by the Kennard-Stone procedure, see Section 5.2, were used as a test set. PLS regression was carried out for the 169 samples and the results are presented in Figure 1(c). Here we get the squared correlation of $R^2 = 0.9980$ and the standard deviation of the residual deviation of s = 0.168 mmol/L. The results of this analysis were applied to the 40 excluded samples and the results are shown in Figure 1d. Here the squared correlation is $R^2 = 0.9974$ and the standard deviation of the residuals equal s = 0.207 mmol/L. Note, that completely new PLS regressions were carried out for each of the 10 groups at cross-validation and at the test set. The use of test set here is not a blind test as recommended by [5] , because all 209 samples

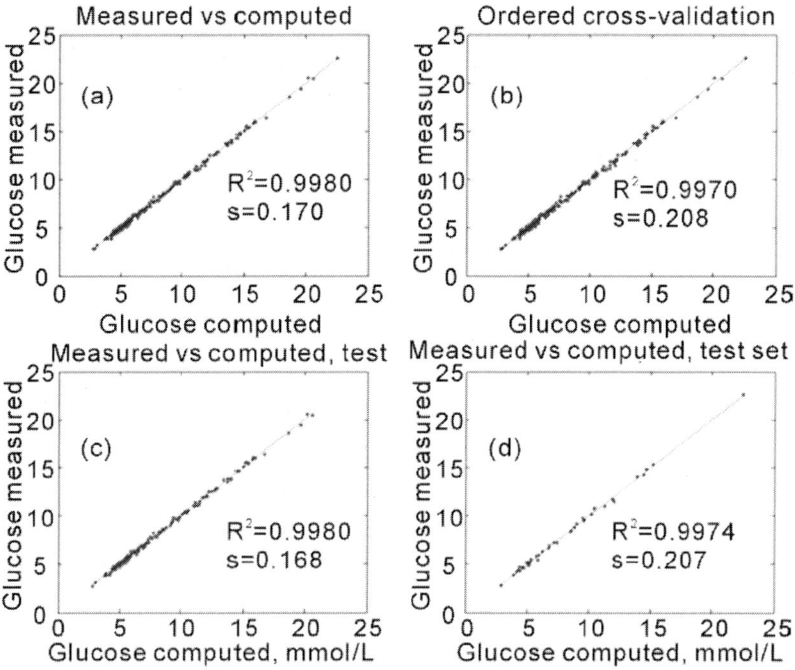

Figure 1: Scatter plots of measured versus computed glucose values.

were used in finding the 50 variables. But a separate blind test gave close to the same results. The results of the analysis are reported by the results from the ordered cross-validation. Thus, we expect to be able to measure glucose by the FTIR instrument with a standard deviation of 0.21 mmol/L. This is satisfactory, because we may expect a standard deviation of 0.12 mmol/L by the laboratory and, as a rule of thumb, one may expect uncertainty of 1% –1.5% by the FTIR instrument. This result of determining glucose values from an FTIR instrument is better than seen in the literature; see [16] and the references there.

Ridge Regression

When there are many variables, it is common to apply Ridge Regression, RR. There is not a problem in determining the inverse for the present data. The condition number is $\lambda_{max}/\lambda_{min} = 396$, where the λ's are the singular values of Singular Value Decomposition of scaled X. The regularized solution of RR is given by $b=\left(X^TX+kI\right)^{-1}X^Ty$, where k is some small constant and I the identity matrix. The theory states that there is a constant k that improves the mean squared error of prediction compared to the least squares solution. There are some theoretical considerations of how to determine k. But they are rarely applied, because the value of k obtained is usually bad. Instead it is common to use the leave-one-out approach. The value of k is used that gives the lowest leave-one-out value, $\Sigma_1^N\left(y_i-\hat{y}_{(i)}\right)^2$, where $\hat{Y}_{(i)}$ is the estimated y_i-value, when the ith sample is excluded. An easy and numerically precise way to carry out this analysis is to use the SVD decomposition of $X_{(i)}$, where i[th] row is excluded from X. The value of k can be determined in two steps. At first the interval (0, 1) is divided into 100 values, 0, 0.01, 0.02, ... The value of k is determined that gives the smallest value. Suppose it is 0.01. Then, secondly, the interval 0 to 0.02 is divided into 200 parts and 200 leave-one-out analyses are carried out. When these have been carried out for the 50 variables, the result is k = 0.0035.

RR analysis was repeated for the 169 samples, where the 40 samples from the Kennard-Stone procedure were excluded. Here the value of k is k = 0.0060.

The RR analysis is be carried out by the algorithm described in Section 3.2 with the weights (11). The value of k in both cases is so small that there is practically no difference from the standard case, where k = 0, which corresponds to PLS regression. In the RR analysis the result is that dimension should be 12. Model control is carried out in the same ways as for PLS regression. Application of the methods in Section 5.3 shows that there are no significant covariances left after dimension 12. The score vectors can be computed by the equation T = XV. Also for the test set. Here the score vectors are not orthogonal. Score vectors from ordered cross-validation, T_c, can also be computed. All plots of score vectors, T, T_t, T_c, against the respective y-values show that there are no correlations after dimension 12. Therefore, one should not compute the full rank solution, but suffice with the solution at dimension 12. In Table 1 is shown the results of the full rank RR solution. The differences between the results are so small that they have no practical value. Therefore, Figure 1 is not shown for RR results.

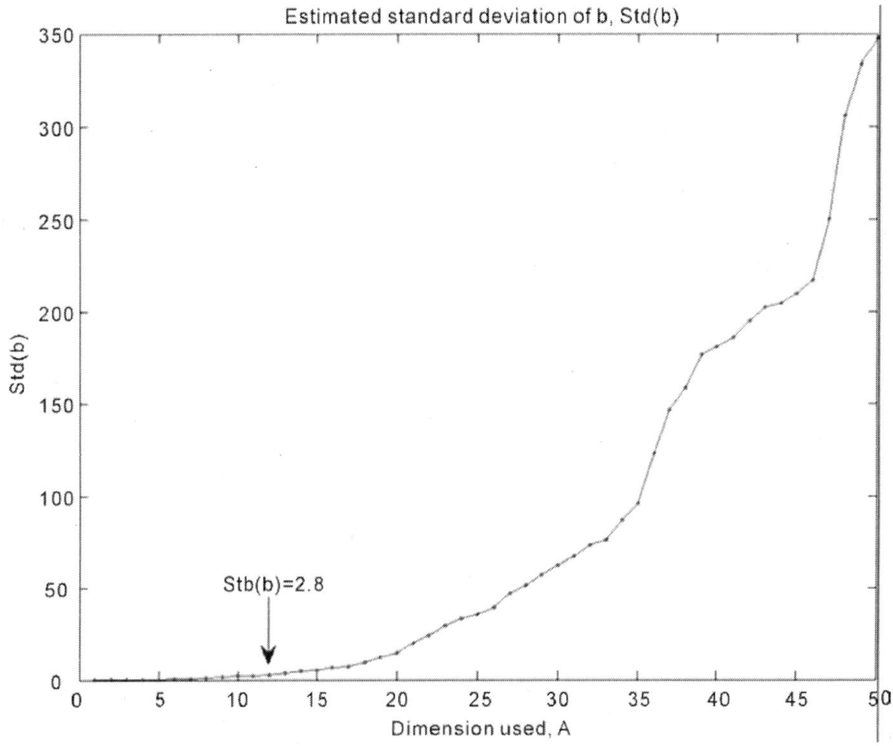

Figure 2: Plot of Std(b) versus dimension, A.

There is a property of RR, which makes RR unappealing to use for scientists. Suppose that there are more samples than variables, N > K. Then, see [1] , for a given k there is matrix Z so that $(X + Z)^T (X + Z) = X^TX + kI$. Thus, RR can be viewed as adding some small random numbers to X. This change of X is not necessary, because there will be a low rank solution for X, which is equally good or better.

In conclusion, only dimension 12 should be used for the RR analysis. In this case there is no practical difference between PLS regression and RR. Full rank RR solution is also almost the same as the PLS one.

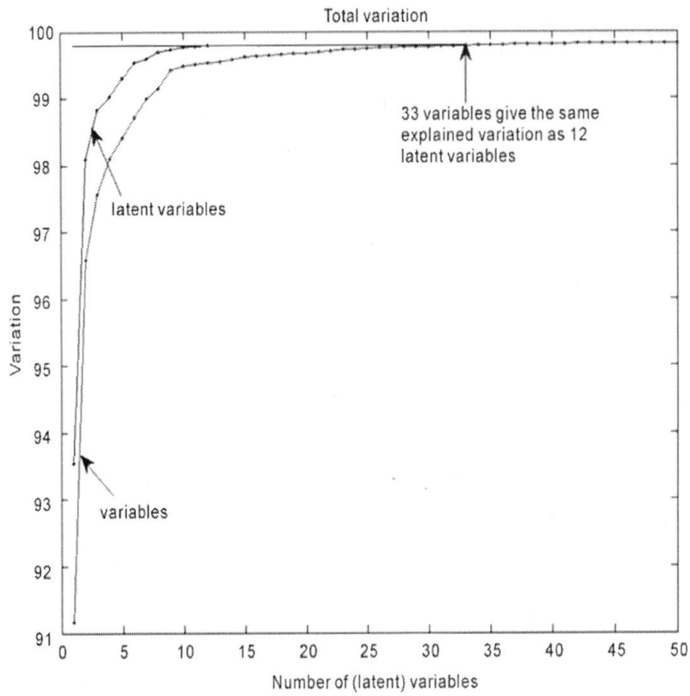

Figure 3: Explained variation, y-axis, versus the number of variables in the model, x-axis. Latent variables start at explained variation of 0.9354 and at the 12th the total variation has reached 0.9980. The variables start at 0.9069 and it is first at the 33th variable that the explained variation of 0.9980 is reached.

Table 1: Results of PLS regression and Ridge regression

		All data	Ord. Cr-val	Reduced	Test
Figure		1a	1b	1c	1d
PLS regression	R^2	0.9980	0.9970	0.9980	0.9974
Dimension 12	s	0.1700	0.2080	0.1680	0.2070
Ridge Regression	R^2	0.9981	0.9970	0.9980	0.9971
k = 0.0035, k = 0.0060	s	0.1640	0.2060	0.1650	0.2170

The Variance Var(b)

There is a common agreement that Var(b) is an important measure. Therefore, it is useful to look at how it increases by the number of dimensions used. This is shown in Figure 2. The trace function, tr, gives the sum of diagonal elements of a matrix. Thus, tr(Var(b)) is the sum of the variance of the regression coefficients. In a plot it is easier to use the square root to make the large values smaller, $\text{Std}(b)=\left(\text{tr}\left(\text{Var}(b)\right)\right)^{1/2}$. For PLS regression it is

$$\text{Std}(\boldsymbol{b})_A = \sqrt{\left[|y|^2 - \Sigma\left(\boldsymbol{y}^T\boldsymbol{t}_a\right)^2 / |\boldsymbol{t}_a|^2\right] \times \left[\Sigma d_a^2 |\boldsymbol{v}_a|^2\right] / (N - A - 1)}$$

It is shown in Figure 2 for A = 1 to 50. At A = 12 the value is Std(b) = 2.8. In a full rank solution the value is Std(b) = 350. A high price is being paid, if full rank solution is chosen. It should be emphasized that Std(b)_A is only a guidance measure. The variance of the regression coefficients can be derived by bootstrapping methods. But the values give guidance for what can be expected. Var(b) is slightly different for RR and the curve is lower (highest value is 115).

VARIABLES AND LATENT VARIABLES

Latent variables are estimates of the main axis in the hyper-ellipsoid, where data typically are located. Latent variables usually pick up the total variation much quicker than the variables. This is illustrated in Figure 3. The Mallows C_p theory is valid for regression analysis both for variables and latent variables.

Latent variables start at explained variation of 0.9354 and at the 12th the total variation has reached 0.9980. The variables start at 0.9069 and it is first at the 33th variable that the explained variation of 0.9980 is reached.

It states that, from a modelling point of view, it is important to keep the size of dimension as low as possible. Typically, the number is much smaller for latent structure regression than for regression methods based on variables.

Scientists/experimenters often tend to ask the question: Which variables are most important? When working with latent structure regression it is difficult to give a precise answer. For the FTIR data we can say that there are 1115 variables. Forward and backward analysis reduces this number to 50. Latent structure regression uses 12 dimensions in the regression analysis. All 50 variables contribute to the regression analysis. It is possible to explain the cumulative effects like at Figure 3.

CONCLUSIONS

A short review of standard regression analysis used in the popular program packages has been presented. It is pointed out that it is a serious problem in industry and applied sciences that the obtained results and presented in papers and reports may not be reliable.

A general framework is presented for linear regression, where the score vectors are orthogonal. Any set of weight vectors can be used that do not give score/loading vectors of zero size. Using the framework, different types of regression analysis can be tailored to different situations. The same type of graphic analysis can be carried out for any type of regression analysis within this framework; also when the start of the computations is variance and covariance matrices.

The algorithm can be viewed as an approximation to the full rank solution. Modelling stops, if further steps are not supported by data. Dimension measures, cross-validation and test sets are used to identify the dimension properly.

The H-principle is formulated in close analogy to the Heisenberg Uncertainty Inequality. It suggests that modelling should be carried out in steps, where at each step an optimal balance between the fit and associated precision should be obtained. A collection of methods, the H-methods, have been developed to implement the H-principle in different contexts.

H-methods have been applied for a number of years to different types of data within applied sciences and industry. The general experience is that they provide better predictions than other methods in the literature.

ACKNOWLEDGEMENTS

The Sime project is carried out in cooperation with Clinical Biochemistry, Holbæk Hospital, Denmark, on determining substances in blood plasma by FTIR spectroscopy. I appreciate the use of the results from the glucose analysis in the present article.

REFERENCES

1. Höskuldsson, A. (1996) Prediction Methods in Science and Technology. Vol. 1, Thor Publishing, Copenhagen.
2. Höskuldsson, A. (2009) Modelling Procedures for Directed Network of Data Blocks. Chemometrics and Intelligent Laboratory Systems, 97, 3-10.http://dx.doi.org/10.1016/j.chemolab.2008.09.002
3. Höskuldsson, A. (2008) H-Methods in Applied Sciences. Journal of Chemometrics, 22, 150-177. http://dx.doi.org/10.1002/cem.1131
4. Höskuldsson, A. (1994) Data Analysis, Matrix Decompositions and Generalised Inverse. SIAM Journal on Scientific Computing, 15, 239-262. http://dx.doi.org/10.1137/0915018
5. Clinical and Laboratory Standards Institute. http://www.clsi.org
6. Kennard, R.W. and Stone, L.A. (1969) Computer Aided Design of Experiment. Technometrics, 11, 43-64. http://dx.doi.org/10.1080/00401706.1969.10490666
7. Siotani, M., Hayakawa, T. and Fujikoshi, Y. (1985) Modern Multivariate Analysis: A Graduate Course and Handbook. American Science Press, Columbus.
8. Roger, J.M., Palagos, B., Bertrand, D. and Fernandez-Ahumada, E. (2011) CovSel: Variable Selection for Highly Multivariate and Multi-Response Calibration. Application to IR Spectroscopy. Chemometrics and Intelligent Laboratory Systems, 106, 216-223.http://dx.doi.org/10.1016/j.chemolab.2010.10.003
9. Höskuldsson, A. (2001) Variable and Subset Selection in PLS Regression. Chemometrics and Intelligent Laboratory Systems, 557, 23-38. http://dx.doi.org/10.1016/S0169-7439(00)00113-1
10. Reinikainen, S.-P. and Höskuldsson, A. (2003) COVPROC Method: Strategy in Modeling Dynamic Systems. Journal of Chemometrics, 17, 130-139.http://dx.doi.org/10.1002/cem.770
11. Grove, H., et al. (2008) Combination of Statistical Approaches for Analysis of 2-DE Data Gives Complementary Results. Proteome Research, 7, 5119-5124. http://dx.doi.org/10.1021/pr800424c
12. McLeod, G., et al. (2009) A Comparison of Variate Pre-Selection Methods for Use in Partial Least Squares Regression: A Case Study on NIR Spectroscopy Applied to Monitoring Beer Fermentation. Journal of Food Engineering, 90, 300-307.http://dx.doi.org/10.1016/j.jfoodeng.2008.06.037

13. Tapp, H.S., et al. (2012) Evaluation of Multiple Variate Methods from a Biological Perspective: A Nutrigenomics Case Study. Genes & Nutrition, 7, 387-397.http://dx.doi.org/10.1007/s12263-012-0288-4
14. Bruker Optics, Germany. http://www.bruker.de
15. Micro-Biolytics, Germany. http://www.micro-biolytics.de
16. Perez-Guaita, D. et al. (2013) Modified Locally Weighted—Partial Least Squares Regression, Improving Clinical Predictions from Infrared Spectra of Human Serum Samples. Talanta, 170, 368-375. http://dx.doi.org/10.1016/j.talanta.2013.01.035

CITATION

1. Höskuldsson, A. (2014) Latent Structure Linear Regression. Applied Mathematics, 5, 808-823. doi:10.4236/am.2014.55077.

Index

C

Centrosymmetric matrix 65, 70
Computer Algebra Systems (CAS) 69
Cross-validation 205, 206, 208, 209, 212, 213, 214, 218

D

DETGTRI algorithm 69, 74
Dynamic way 28

E

Exchange matrix 62, 63, 64

F

Full orthogonalization method (FOM) 19, 20, 26, 27, 32, 42, 44, 45

G

Gamma function 177
Gamma matrix function 178
Gaussian elimination (GE) 78
Global lower 4

H

H-principle 190
Hypergeometric matrix function 173, 174, 177, 178, 179, 180, 183

I

Important application 154
Independent variables 191
Integral representation 173, 181, 182, 183, 184, 185, 186

L

Linear map 47, 54, 55, 56, 57, 59
Linear regression 189, 190, 191, 193, 196, 201, 218
Linear system 1, 2, 3, 4, 7, 14, 17, 19, 20, 21, 22, 25, 34, 42, 109
Linear systems 61, 62, 75, 77, 78, 80, 81

M

MAPLE 61, 62, 64, 69, 81, 83
Matrix theory 47
Multiplicative structure 153, 154, 155

N

Notation. matrices 191
Numerical experiment 1, 3, 14

P

Principal Component Regression (PCR) 194

Q

Qualitative properties 174

R

Recurrence relation 173, 174, 182, 186
Regression methods 189, 190, 194, 217

S

Square matrix 21, 25
Square root 38
Standard regression analysis 189, 218
Sturm sequence 19, 21, 29, 30

T

Triangular part 110, 112, 115